# *Frontiers of Space, Time, and Thought:*

*Essays and Stories*

*on The Big Questions*

Books by Edward M. Lerner

*Probe* (*)
*Moonstruck* (*)
*Creative Destruction* (collection)
*Fools' Experiments*
*Small Miracles*
*InterstellarNet: Origins* (**)
*InterstellarNet: New Order* (**)
*Countdown to Armageddon / A Stranger in Paradise* (collection)
*Frontiers of Space, Time, and Thought* (collection) (**)
*Energized*

Books with Larry Niven
*Fleet of Worlds*
*Juggler of Worlds*
*Destroyer of Worlds*
*Betrayer of Worlds*
*Fate of Worlds*

* Reissued by FoxAcre Press
** Published by FoxAcre Press

# *Frontiers of Space, Time, and Thought:*

## *Essays and Stories on The Big Questions*

Edward M. Lerner

*Frontiers of Space, Time, and Thought: Essays and Stories on The Big Questions*


Print Edition ISBN 1-936-77137-3
Print Edition ISBN-13 978-1-936771-37-0



Visit the author's website at
www.edwardmlerner.com

**Fictional contents, in order of original publication:**

"Unplanned-for Flying Object" first appeared in *Analog Science Fiction and Fact* (September 1994)

"Grandpa?" first appeared in *Analog Science Fiction and Fact* (July/August 2001)

"A Matter of Perspective" first appeared in *Artemis, Science Fiction for a Space-Faring Age* (Winter, 2003)

"Great Minds" first appeared in *Jim Baen's Universe* (October 2006)

"RSVP" first appeared in *Darker Matter* (March 2007)

"Chance of Storms" first appeared in *Jim Baen's Universe* (April 2007)

"At the Watering Hole" first appeared in *Jim Baen's Universe* (August 2007)

"Inside the Box" first appeared in *Asimov's Science Fiction* (February 2008)

"Where Credit Is Due" first appeared in *Analog Science Fiction and Fact* (October, 2008)

"Small Business" first appeared in *Analog Science Fiction and Fact* (January/February, 2009)

"No GUTs, No Glory" first appeared in *Jim Baen's Universe* (August, 2009)

"A Time for Heroes" first appeared in *Analog Science Fiction and Fact* (June, 2010)

"Blessed Are the Bleak" first appeared in *Analog Science Fiction and Fact* (April, 2011)

**Nonfiction contents, in order of original publication:**

"Beyond This Point Be RFIDs" first appeared in *Analog Science Fiction and Fact* (September 2007)

"Follow the Nanobrick Road" first appeared in *Analog Science Fiction and Fact* (September 2008)

"Insignificance" first appeared in *Analog Science Fiction and Fact* (October, 2009)

"Rock! Bye-Bye, Baby" first appeared in *Analog Science Fiction and Fact* (November, 2009)

"Say, What? Ruminations about Language, Communications, and Science Fiction" first appeared in *Analog Science Fiction and Fact* (March, 2011)

"Lost In Space? Follow the Money" first appeared in *Analog Science Fiction and Fact* (October, 2011)

"Faster than a Speeding Photon" first appeared in *Analog Science Fiction and Fact* (January/February 2012)

# Contents

# *Foreword: The Big Questions*

The universe is a really big place.

How big? No one knows, of course. The universe is so immense that—even in principle—parts of it are too distant to be seen.

Astronomers measure the observable universe in light-years (billions of light-years, actually), because familiar units of distance are so inadequate to the task. In a year, light in a vacuum travels about ten trillion kilometers, or six trillion miles, not that our brains are wired to think in trillions. (We wouldn't so blithely run up debts in the trillions if we did—one person's opinion.) By way of comparison, NASA's space probe *Voyager 1*—the farthest-ranging human-built object, on its travels since 1977—has crossed a few billion kilometers. Not that our brains do well with billions, either ....

And the universe is old, with an estimated age of about 13.5 billion years.

And perhaps what most of us think of as *the* universe is merely one inconsequential portion of an immense multiverse.

That's looking outward. Peering inward, physicists find things as amazingly tiny as the universe around us is astonishingly large. Each of us is comprised of trillions of atoms. Atoms are measured in the trillionths of a meter—and yet each atom, comprised of yet tinier elementary particles, is mostly empty space. Our brains are no more wired to contemplate subatomic particles, and the quantum-mechanical weirdness that describes them, than we are to grasp light-years.

So: a vast universe around us. An unseen vastness within. Eons of time. It's enough to make a person wonder about things. I know I do.

Are we alone in the universe? If not, could we manage to communicate with aliens? What hope is there of ever exploring the immensity of the universe? And because every journey begins with a first step … will humanity take that first step and leave our planetary cradle?

I wonder about slightly less cosmic matters, too. Can privacy survive the Internet? What will the Internet even look like a few years from now?

I've been a physicist, a computer scientist, a writer of science and technology articles, and a science-fiction author. Along the way, I've visited major observatories, flown the space-shuttle training simulator, toured a satellite factory and a semiconductor foundry, hobnobbed with the experts about how best to deflect an onrushing asteroid—and done lots more mind-expanding stuff. More often than not, I've written about it.

This book collects science and technology articles I've written on some of these Big Questions, together with more than a dozen short stories touching on the same topics. With one exception, all are collected here for the first time. ("Small Business" appears in an earlier collection, *Countdown to Armageddon / A Stranger in Paradise*. The story fits so well with two of the essays that I couldn't bring myself to omit it.) Because of my, shall we say, eclectic background, the essays often point to examples from science fiction where particular points of science and technology are interestingly illustrated. A few pointers reference my own writing—it helps to know *exactly* what the author had in mind—but more typically not.

I've lightly edited the essays from their original magazine layouts for format consistency. I've taken out the occasional aside to frequent readers of those magazines. The essays originally appeared in print, but as time goes by fewer and fewer readers will experience this collection that way, and footnotes that are convenient on the printed page are nuisances on e-readers. (*Today's* e-readers, anyway.) So: I've converted many former footnotes into inline text and (in the ebook editions) the notes that remained were switched into per-essay end notes.

Most notes are URLs that lead to source materials or related discussions. All links are confirmed active as I write (occasionally after I made substitutions for vanished pages)—and by the time you read this, doubtless some have gone stale.. That's the web for you.

I've also selectively updated the essays, the earliest of which saw print in 2007. Not every short-term prediction and forecast has come true, although many have. Judge for yourself my prognostication skills—I've clearly identified the scattered content changes I've made to reflect the latest science and technology.

And while you read, I have more thinking to do....

*—Edward M. Lerner*
*February, 2012*

# *Are We Alone?*
# *(Life, the Universe, and Everyone)*

Our first Big Question: in all the vastness of space and time, are we humans alone? Either answer would be profound.

The data are more encouraging than ever that humanity *might* share the universe with other life.

Only a few years ago, ours was the only known planetary system. The Extrasolar Planets Encyclopaedia[1] now lists hundreds of other planetary systems, the numbers growing more or less daily. As instruments and search techniques improve, astronomers find smaller and smaller planets. The discovery of extrasolar planets much like Earth is expected within the next few years.

Everywhere biologists look, Earth teems with life: deep within the rocks, in boiling-hot ocean-floor vents, in cooling ponds filled with the radioactive waste from nuclear power plants, and afloat high in the atmosphere. It appears that life—wherever it finds a foothold—evolves and prospers.

Still, none of that is proof of extraterrestrial life. Evolution explains the richness of life but not its origin. The great leap from lifelessness to life remains unexplained by science; we simply don't know how often that transition occurs.[2]

Suppose that somewhere (the still hypothetical) alien life, like Earthly life, evolved all the way to intelligence. In biological terms,

1. http://exoplanet.eu/catalog.php

2. *The Eerie Silence: Renewing Our Search for Alien Intelligence*, Paul Davies, includes an excellent discussion of the puzzle of biogenesis.

people and the aliens would have less in common than people and paramecia. That leads to another Big Question: could humans and another intelligent species—if and when we encounter one another—find a way to communicate?

This section of the collection opens with "Say What? Ruminations about Language, Communications, and Science Fiction." That essay is followed by three short fictional takes on the First Contact theme.

# *Say What? Ruminations about Language, Communications, and Science Fiction*

## (First published in March 2011)

Vanguard's *bridge was a lonely and boring place, but as a matter of tradition watch officer Joan Miller served her shift without complaint. To come upon more than a mote of dust or fleck of ice here in the depths of interstellar space would make the shift eventful indeed.*

*Only suddenly something* was *out here, and it was no mere fleck of ice.*

*Eyes wide, Joan studied her sensor array. Gravity waves had drawn her attention. Something was making those waves. Something massive, moving fast, exploiting technology far beyond that of Earth. Something heading straight at her.*

*A* ship*?*

*Her heart pounded. At long last, humanity might have found a companion intelligence.*

*Before Joan could decide how to share the monumental discovery with her shipmates, an LED flickered on the comm console. With a trembling hand, she accepted the hail. Her holo tank filled with what brought to Joan's mind the crossing of a walrus with a lobster. It wore an ornate garment of some sort, replete with sash, braid, buttons, medals, and gold epaulets upon its (four) shoulders.*

*"I say," the creature began, its chitinous mandibles sliding over one another, its brush-like mustache wriggling, "Jolly good show meeting here, eh wot?"*

◆❖◆

Threw you right out of the story, didn't I?

Why does the walrus/lobster know English, let alone speak English like a refugee from a Gilbert and Sullivan operetta? With chitinous mandibles, how does it even make sounds reminiscent of human speech?

We'll chalk up that dreadful story snippet to making a point. To wit: in science fiction, language and communication details matter. In this essay we'll look at how language and communications can enrich an SF story—and at ways to bypass the related complications when communications details are less than central to the tale. (Our topic goes beyond human-alien encounters. As we'll see, languages can change over time. Stories set in any era besides our own [including, but not limited to time-travel stories] should consider language issues, too. Ditto alternate-universe stories, even those set on a parallel "Earth.")

Let's begin by considering how different from English even human languages can get.

### *Variability in human languages: the view from low Earth orbit*

This article can't begin to do justice to the range of differences among human languages.[1] If you've studied any second language (or English is your second language), you already have a window into the ways that humans differ over the nature of languages. Without exhausting the range of variations from English among human languages, consider:

- How many tenses should there be? Spanish has two past tenses, one for events that definitely occurred once and another for events that repeat or continue.

1. For a broader survey of human languages and an alternate take on alien speech, see the chapter "Alien Language" in *Aliens and Alien Societies: A writer's guide to creating extraterrestrial life forms*, by Stanley Schmidt.

- What's the domain of gender? French and German assign genders to inanimate objects.[2]
- How many pronouns are needed? English, since it (mostly) eliminated *thee* and *thou*, has nearly eliminated the distinction between familiar and formal pronouns. The exception—most evident, one supposes, in the UK—is the royal we.
- What building-block sounds comprise a language? English lacks the guttural "ch" of Scottish, German, and Hebrew. Western languages lack the clicks of southern African languages like Xhosa and Zulu.
- How should languages be symbolized? Humans haven't agreed, with billions of us embracing alphabetic systems and billions more using ideographs and syllabaries.

With such variability among human languages, how likely is our walrus/lobster friend to speak English?

### *Let's get physical*

Humans evolved to communicate via modulated sound waves. Fair enough: the noises we make can cross moderately long distances through the atmosphere, are reasonably non-directional, and can encode complex messages. But what other methods might have worked?

It requires no great stretch of imagination to suppose communication by modulating emitted (or reflected) light. Terrestrial life offers early steps in that direction. Pit vipers have infrared sensors apart from their eyes.[3] Octopi and squids camouflage themselves and signal emotion through changes in skin color.[4]

---

2. For much more about gender differences among languages, see "Der Mann, Die Frau, Das Kind," by Henry Honken, in the June, 2010 issue of *Analog*.

3. http://www.physorg.com/news76249412.html

4. http://www.scientificamerican.com/article.cfm?id=how-do-squid-and-octopuse

And why limit ourselves to visible and near-visible light when the electromagnetic spectrum is so broad? Aquatic life as varied as sharks, lungfish, and catfish have rudimentary electric-field sensors.[5,6] Many migratory birds orient themselves with magnetic sensors.[7]

Other terrestrial species communicate by gesture (the dance of bees[8]) and chemically (ants leaving pheromone trails to guide foraging expeditions[9]). Human sign languages demonstrate that a gesture-based language can convey information in as nuanced a manner as voice. As one example among many, consider American Sign Language.[10] And while chemicals released into wind or water seem an unreliable means of communication, chemical packets physically delivered to a receptor—no different, in principle, than handing someone a written note—could be quite reliable.

It's far from certain that language-capable aliens will use sound waves for their communications.

### *Picture this*

Humans evolved to communicate with sounds, but (mostly) we perceive the world visually. "I see" is synonymous for "I understand." How different would languages be if communication and primary perception shared a medium?

---

5. http://www.absoluteastronomy.com/topics/Electroreception.

6. The electrical capabilities of some marine life go beyond sensing to communication, navigation, and even stunning of prey. See: http://www.answers.com/topic/electric-organ-biology

7. http://news.nationalgeographic.com/news/2007/09/070927-magnetic-birds.html

8. http://www.cals.ncsu.edu/entomology/apiculture/pdfs/1.11%20copy.pdf

9. http://icouzin.princeton.edu/pheromone-trail-networks-in-ants

10. http://en.wikipedia.org/wiki/Asl

Consider dolphins and their sonar-based navigation. Their sonic pings return a 3-D representation of the nearby ocean. Can dolphins imitate those echoes? If so, they can directly communicate 3-D images to their pod mates (as do, for example, the enhanced dolphins in David Brin's Uplift universe). And if dolphins can emit sonic images of real scenes, why not also sonic images of imagined items?

It's not a big step from imagining sonar-imaging dolphins in Earth's oceans to radar-imaging aliens on another planet.

A familiar adage has it that a picture is worth a thousand words. Certainly human languages require many words to describe a visual scene. Our languages would be quite different if we could directly "paint" pictures.

### *Where am I?*

Languages encapsulate, among other things, our understanding of the world and our relationship to it. How might languages differ if our sense of the physical world differed?

Proprioception is the sense by which we relate the position of our body parts to each other and to the external world.[11] Humans are bilaterally symmetric, with sensors that favor a particular direction (i.e., define "forward").[12] Reflective of our physiology and proprioception, our speech is rife with references to that which is in front of, next to, and behind us. We place quite different values on what transpires to our faces and behind our backs.

Creatures with body plans other than ours might perceive—and describe—the world quite differently than humans do. Imagine a trilaterally symmetric alien, with limbs and sensory clusters spaced every 120° around its body. It walks, reaches, and senses equally well in any direction. Front/back and left/right distinctions do not

---

11. http://en.wikipedia.org/wiki/Proprioception

12. Most of us aren't even quite bilaterally symmetric—we're left or right handed. And so (righties being chauvinists), English offers us *gauche* (from the French word for left), as a synonym for tasteless and vulgar, and *sinister* (from the Latin word for left) as a synonym for threatening and creepy.

apply. *No* reference solely to its body suffices to locate an object—or a fellow alien—relative to itself.

In my novel *Moonstruck*, the Krulirim have this body plan. I had to give these aliens a magnetic sense. A Krul locates an object in part by reference to the angle between a line toward the thing and a line toward the nearest magnetic pole.

With my left/right, front/back relationship to the world, it's natural for me to locate an item as (for example) "three meters in front of me and four meters to my right." A Krul (ignoring its use of meters and degrees as units of measure) might say in that circumstance that the item is "five meters distant from me and in a direction thirty degrees east from my bearing to the North Pole."

Now imagine the Krul relating the relative positions of separated objects or describing those objects' positions to another Krul. The Krul's propriocentric sense—and so, its language—necessarily deals with trigonometry.

Consider group minds like, but more intelligent than, the bees of a hive.[13] What sort of proprioception—if any—might the collective mind have? How might its propioceptive "read-out" change as the collective's members disperse and re-gather? How would one such collective mind describe the position of an object relative to itself? To another collective mind?

### *A moving target*

Languages evolve.

Vocabulary changes. We don't exactly speak like characters in a Shakespeare play. ("Who would fardels bear?" Hamlet asks in his famous soliloquy. A *fardel*, it turns out, is a Middle English term for a burden.) Consider end-of-year "top word" lists (a couple of 2009 examples: H1N1 and [as a verb denoting concise communication] twitter). Consider the flood of terms entering the language from:

- Company names (like Xeroxing and Googling).
- Acronyms (like RN, UFO, SAT, and AIDS).

---

13. Or less extreme than an intelligent hive, the Tines: intelligent packs of dog-like creatures in Vernor Vinge's *A Fire Upon the Deep*.

- Commercials and popular culture (like "Where's the beef?")
- Science and technology (everything from a veritable bestiary of subatomic particles[14] to such neologisms as dark matter, blog, and carbon footprint).
- Current events (such as anything with the –gate suffix, from the name of a hotel that saw "a third-rate burglary attempt").

Expressions go in and out of vogue. (Is "out of vogue" yet out of vogue?) "Hold your horses" is, to be charitable, on the dated side. "Too big to fail" is a recent coinage reflecting events that we would all be happier not to have experienced. Verbs become nouns, and vice versa, such that dynamite and telephone can be either part of speech.[15], [16] Idioms borrow from literature (say, Catch-22) and from mythology (say, Pandora's Box) in ways that don't translate.

Pronunciations change, too, sometimes for no better reason than that people can't always be bothered to enunciate clearly. English is full of anachronistic spellings: time capsules. And so it is that "rough" rhymes with "tough," but neither rhymes with "though."

Languages fragment and diverge. For example, around 500 B.C. Proto-Germanic split off from Indo-European. About a millennium later, one of Proto-Germanic's many offshoots split into High and Low German—high and low being geographic distinctions—on the figurative road to modern German and Dutch (and amid that split, pronunciations shifted, too.) The fallen Western Roman Empire's onetime European territories evolved various "Romance" languages (such as French, Italian, Spanish, Portuguese, and Catalan) by blending Latin with the local vernaculars—local vernacular be-

---

14. Even subatomic particles can have nicknames. The Higgs boson—the particle that, in theory, provides matter with the attribute of mass—is sometimes referred to as the "God particle."

15. As a *Calvin and Hobbes* cartoon would have it, "Verbing weirds language."

16. Dynamite can even be an adjective, as in: that was a dynamite party last night.

ing variable, too, as Goths, Huns, Franks, and other ethnic groups shifted about Europe.

Languages collide. Redundant terms speak to successive conquests of England. Think house (from the German "haus") and (upscale, from Old French, because the Normans were the last to conquer England) mansion.

Languages borrow from one another. You can travel Europe ordering a hamburger (in English, French, German, and Italian), hamburguesa (in Spanish), or hampurilainen (in Finnish). (As much as I enjoy hamburgers, I don't recommend this course of culinary action.)

National authorities also sometimes try (with mixed success) to resist such borrowings. Thus the "English" neologism *telephone* (from Greek roots *tele* [far] and *phone* [sound]) gave way for a time, by official decree, to the German neologism *fernsprecher* (far speaker).[17] The English neologism *computer* was rejected in favor of the French neologism *ordinateur*.[18]

How quickly do languages change? Without attempting to quantify change, conversation with adults merely one generation younger or older than I suggests an answer. Very.

---

17. My college German instructor, a grad student from Germany, taught that the Nazis purged many French and English technical terms. I find similar anecdotes on the Web, but nothing definitive—perhaps because I retained too little from that long-ago class to read articles in German. Google Language Tools offers the following German translations for English telephone: Telephon, Telefon, Fernsprecher, Apparat, Telephonapparat, Telefonapparat, and Fernsprechapparat. Germans didn't simply adopt the (classically derived) term from the inventor's English.

18. The French government has long taken an interest in language purity. L'Académie française (http://www.academie-francaise.fr/) was established in 1635 as the official arbiter of the French language. Enforcement of language standards is very real, as GE Medical Systems learned to its dismay in 2006 when a French court ordered translation of all internal technical documents. So note: L'Académie française prefers that you order viande hachée in lieu of that proverbial hamburger.

In short, human languages reflect more than our physiology—they reflect our history, technology, and sociology, too. Aliens' languages may have similarly complicated origins.

### Time out

What about stories set in Earth's future? Humans aren't evolving so quickly we need to worry about changes to future humans' sense organs or body plan.[19] Or we can set our stories in the past to preclude changes to human nature.

As we saw earlier, languages transform over time. Set a story in another era without addressing the language differences ("Chill out, dude," Socrates said) and something is apt to seem amiss.

### Communicating with strangers: the easy cases

If two languages—and the beings who created them—are sufficiently similar, the difficulty can be finessed. After all, people *do* learn new languages. A story that opens long after humans and aliens (or whomever) met can credibly assume that at least one side's linguists previously figured out the other's language.

Translation programs get written even today, although today's state-of-the-art in translation software is as often humorous as helpful. Correct and complete translation requires an understanding of both languages, their historical contexts, idioms, and literary traditions. Translation software, at its core, is a matter of artificial intelligence—and AI is certainly a staple of science fiction.

And if I may digress, also a Very Hard Problem.

How self-absorbed are humans to use the Turing test as the criterion for when artificial intelligence has been achieved? The test: if a human can't tell from a typed dialogue that he's dealing with a

---

19. Or do we? In the not-so-distant future, we might learn to gengineer gills or new senses, enhance our natural senses with implanted prostheses, or upload our minds into (among the possibilities) a virtual reality, robotic shell, or cloned body. For a few of the possibilities, see *Beyond Human: Living with Robots and Cyborgs*, by Gregory Benford and Elisabeth Malartre This being an essay, not a book, let's set aside new-mode humans for now.

program, that program will be deemed intelligent. By the mirror standard, a competent program might judge humans unintelligent.

Looking at human speech—or the purportedly meaningful contents of tweets and IMs—a case could be made that the software was correct.

End of digression.

Stories dealing with travel to the past have a lower language hurdle. (Have you kept track of the metaphors that keep intruding? Pondered whether they're apt to translate?) Our Hero can research the historically appropriate dialect of the language he'll find spoken at his destination before jumping through time. Even then, the time traveler may, and probably should, be surprised once he arrives by idioms and pronunciation shifts absent from the historical record.

So much for the easy cases. What happens when the others' languages, native environments, or worldviews are very different from ours?

### *The cavalry trope to the rescue*

A *linguistic trope*, such as a metaphor or simile, uses words or expressions in a non-literal way. A *literary trope* is a common theme in storytelling, well-known examples including heroes and quests. Assuming a degree of reader familiarity with common literary tropes, authors can avoid spelling out related specifics.

Science fiction, as a genre, has its own literary tropes. Some *science-fictional tropes* use known science in a non-literal way, such as by postulating a faster-than-light space drive, a method of time travel, or a true artificial intelligence.

(AI—at the level of an entity with which we can interact as though with a human—continues to elude us. What passes for AI, such as chess-playing programs and expert systems—remains too disconnected from the physical world to exhibit a human's "common sense," and so continues to fail the Turing test. Consider how different a program's environment and experience are from your own. That difference alone may explain humanity's limited success with developing artificial intelligences.

By comparison to an AI, biological entities native to even a very unearthly world may have an easy time establishing a common ground with us. Such as already having the concept of ground …)

I think of an SF trope as a willing-suspension-of-disbelief contract between author and reader. As with any contract, both parties must agree.[20]

An SF trope works best—in one person's opinion, anyway—in the background, rather than front-and-center in a story. (Try explaining *that* idiom to a Krul.) By way of analogy, suppose wind-powered sailing ships had yet to be invented. A science-fictional *Robinson Crusoe* might treat the ship as a trope. The vessel served its purpose in the story when it sank; the mechanics of sailing don't matter. In contrast, consider science-fictional versions of *Moby Dick* or *Two Years Before the Mast*. Those stories can hardly be told without showing the technology of sailing ships.

It's not 100% clear-cut when a story premise is a trope. Science-fictional time machines are usually tropes, ways to shift a modern character—someone with whom the reader will empathize—to the time(s) of the story. But when I encounter a passage such as:

> **"Early in the 22nd century, physicists succeeded in dependably stabilizing entangled quantum particles. When the entangled particles were tachyons, quantum teleportation became time travel …"**

I infer that the story is intended as hard SF. (Hard SF: stories based on hard—that is, real—science, not stories that are difficult to read. Hard SF avoids contradicting what we (think we) understand about the way the universe works.) In the above story snippet, the nature of the supposed time-travel mechanism and its underlying constraints may prove to be central to the story.

We now return this essay to matters of communications …

---

20. Some readers—and some authors—won't sign that contract. See the November 2009 *Analog* editorial "Aiming High—or Low."

### *Language-related tropes*

SF authors use many tropes to circumvent the problem of human/alien communications.

The simplest science-fictional language trope, so deeply embedded as to go unnoticed, is the *universal language.* It's not that everyone suddenly adopts Esperanto.[21] Rather, future (or alien) science has discovered principles that—unknown to human science today—underlie all communications. Anyone knowledgeable of the underlying principles of the universal language can readily master any other language.

Does the universal language exist? No one has disproven it, so its use remains fair game for SF. That said, it's difficult to imagine what underlying principles could encompass the many historical and physiological differences between human and all alien languages.

Humans, with the advantage of a common physiology, have yet to achieve a global language. [22, 23] Technological shortcomings might once have precluded convergence on a single human language, but with radio, television, the Internet, and on-demand publishing, we could—at least among the high-tech societies—standardize on a single language. I don't foresee that happening anytime soon. That begs the question: how many languages are apt to be extant within an alien civilization? If only one, that discovery alone would suggest interesting cultural differences between the aliens and humanity.

If a universal language exists, that's no guarantee everyone is physically able to understand or speak it. One may still need a *univer-*

21. A language designed to be neutral (among political and ethnic groups) and easy to learn. Despite its laudable goals, few people (and no countries) have adopted it. http://en.wikipedia.org/wiki/Esperanto

22. (Or we had a global language and lost it, if you take literally the Tower of Babel story.)

23. If George Bernard Shaw is to be believed, even a shared language has its limits. To wit: "England and America are two countries divided by a common language."

*sal translator* to perform the translations, or to convert between such disparate formats as modulated sound waves and aroma blends. The universal translation device is only as plausible as the language skills and technological mechanisms by which it has been implemented.

The *Star Trek* franchise makes frequent use of a universal translator—or a not-quite-universal translator when plot logic requires a bit of misunderstanding. Viewers accept the (il)logic because a new linguistic puzzle every week would get tedious.

Perhaps we or the aliens can bypass the incompatible-language barrier through *telepathy.* Setting aside questions of how—even at the handwavium level of a trope—mind reading might work, this hypothesis raises questions about the relationship between thought and language. Unless thought operates independently of language (which doesn't seem to be the case in my head), or thought uses the undiscovered universal language, it's hard to see how telepathy can overcome language barriers. And for telepathy to work across species—especially species with disjoint neural mechanisms and biochemistries—requires an additional leap of faith.

Consider the range of possible data formats. If analog: modulated onto the strength/weakness or prevalence/scarcity of what underlying medium? If digital: using what symbol set, and with what type of error-correcting coding? If neural, which chemical(s) are operative? Is information localized, as in familiar digital storage, or distributed, as in holograms and our own neural wiring? That's only a small sample of the variability we might encounter among independently evolved biospheres.

(And you thought HD-DVD vs. Blu-ray was a hard problem.)

Finally—and not for the faint of heart—we come to the shush!-we've-all-agreed-to-look-the-other-way language trope. The *Stargate: Atlantis* franchise features remote worlds whose populations descended from humans abducted thousands of years earlier. Pegasus Galaxy humans speak English even though their ancestors left Earth before there *was* English. It's an impressive feat that certainly moves along the storyline. Just try not to think about it ….

### *Putting it all into words*

So: your fictional aliens, alternate-Earth residents, and humans from other eras have language(s)—or at a minimum, pronunciation(s) and vocabulary(ies)—different from our own. By technology or by trope (this expression is hereby offered as a replacement for "by hook or by crook"; reasonable terms available) the human or human-empathetic characters establish communications. An important question remains: how, within the story, do we represent that communication? Writers have used many techniques:

- Snippets of foreign vocabulary rendered into a familiar character set sprinkled into otherwise English text. Items without human equivalents, like advanced technologies of the aliens, retain their alien character. (For example, the bigger-than-worlds artifact that is the Ringworld is constructed from a substance so strong and dense that it blocks a substantial fraction of solar neutrinos. That's amazing stuff, without any human equivalent. And so, throughout Larry Niven's Ringworld series, that material goes by the natives' term: *scrith*.)
- As above, but with foreign usages limited to proper names. Guillaume, when he visits London, may speak English, but he doesn't start referring to himself as William. (Of course English readers can pronounce, to greater and lesser degrees, the name Guillaume. What if the alien names are hard or impossible to pronounce? I have it on good authority—read: irate emails—that such words annoy some readers. Should I as an author avoid hard-to-pronounce terms? Perhaps sometimes: when language cues aren't meant to emphasize the alien nature of the nonhumans.)
- Aliens with great language skills, who have mastered English so that the human characters (and readers) don't have to learn, say, Tau Cetian.
- Arcane word orders that make the English *seem* alien. (Think of, in the Star Wars universe, Yoda and Jar Jar Binks. Better

yet, don't think of them. Annoying, they are. Strong, the aggravation becomes. Weirdness, an inferior substitute for alien is. On the fantasy side of speculative fiction, J. R. R. Tolkien used archaic wordings to great effect to give Middle Earth a terrifically *other* feel.)

- Alien speech and writing rendered into non-Roman character sets, even to non-linguistic symbols such as musical notes, with translations nearby. (Such as the Mother-Thing alien sings/speaks in Robert A. Heinlein's *Have Space Suit—Will Travel.*)

The author must fit the technique(s) to the story. And nowhere is that challenge harder than where tropes just won't serve ….

### *Search for Extraterrestrial Intelligence (SETI)*

SETI inherently deals with another civilization's communications. Suppose Earth's observatories hunt for extraterrestrial signals, perhaps transmitted our way as microwaves or laser beams. What modes of long-range communication might the aliens use? Suppose a signal is detected. What method(s) might extract the signal's meaning? If a signal seems intended for us, how will we ascertain what the aliens are attempting to say? If a signal seems accidental—their analogue, perhaps, of *American Idol*, spreading at light speed across the galaxy—what meaning might we manage, through cleverness, to extract?

It's hard to imagine a SETI story that is not at least somewhat concerned with the details of communication. Tropes won't do—there *is* no story without delving into the method by which the signal is received, what within the signal is understandable, or the cleverness with which that meaning is determined. In a SETI story, we can hardly start to understand the aliens by tapping our chests and over-enunciating our names.

How might we find meaning in an alien radio broadcast? It's not as though we can hope to uncover something like the Rosetta Stone.[24]

---

24. An archeological artifact discovered in Rosetta, Egypt, in 1799. Identical passages inscribed in Egyptian and Greek symbols gave valuable clues to reading hieroglyphics. http://en.wikipedia.org/wiki/Rosetta_stone

Except that for some messages *nature* is a Rosetta Stone. To the extent physical laws are universal—and known by both parties—representations of those physical laws may provide a basis for mutual understanding.

Why *may*? Because common understanding of physical principles does not assure compatible representations. Newton and Leibniz, for example, developed quite different representations for differential calculus. Schrödinger's equation and Heisenberg's matrix mechanics are very different representations of quantum mechanics. And on, and on …

Lots of hard work would be required to establish a mutual understanding of the physical domain.

If limited translation succeeds from clues provided by physical laws, what then? It's less than intuitive how to proceed from a specialized vocabulary based upon shared physical discoveries to the broader vocabulary necessary for unrelated-to-science topics. "Catch-22" is far afield from the periodic table.

I'm hardly the first SF writer to wonder how distant species might establish communications. Nor am I the first to suggest that the universality of physical laws would provide common ground. H. Beam Piper's "Omnilingual" did the latter way back in 1957. The story involves astronauts finding the ruins of an extinct Martian civilization and one explorer recognizing a periodic table—and seeing its value in cracking open the Martian library.

(My novelette "Dangling Conversations" opened with a SETI discovery and went on to delve deeply into interstellar message coding issues. Expanded and combined with related stories, "Dangling Conversations" provided the opening segment of my novel *InterstellarNet: Origins*. As the term InterstellarNet suggests, the communications issues get much more sophisticated over time.)

### *A trail of bread crumbs*

Earth's earliest transmissions were likely submerged in the naturally occurring radio emissions of the Sun, but perhaps aliens with sensitive receivers can make out humanity's more recent transmissions. It'd be interesting if visitors were to back-track such signals and come calling already conversant in English (or Cantonese, or whatever).

But can Earth's transmissions teach human languages? Our broadcasts are hardly designed for an alien audience. The transmissions employ a multitude of frequencies (not all of which propagate equally well through the interstellar medium), carrier-wave modulation schemes (e.g., amplitude and frequency modulations), data representations (e.g., audio and audiovisual, analog and digital, "in the clear" and encrypted[25]), and spoken languages. Superimposed over the unending radio and television chatter are powerful radar pulses.

Suppose nearby aliens (including those simply passing through the neighborhood) intercept Earth's transmissions of the past several decades. Reconstructing video from TV broadcasts may be a simpler problem than extracting meaning from radio broadcasts or TV's audio subchannel. If so, imagine what impression video alone—say, from the evening news—might make. Perhaps humanity should hope that "we're here, and we use technology" is the only significance that distant visitors can attribute to our transmissions.

(My Krulirim discovered Earth by back-tracking TV broadcasts. And encountered, amid all that chatter ... *Sesame Street*. Say what you will about Big Bird, his show might be educational for aliens, too.)

### *First Contact*

First Contact stories up the ante: the interaction is face (or whatever) to face. Where SETI stories look for the commonalities between species that might enable communication, First Contact stories more often focus on our differences. It's hard to imagine a First Contact story in which inter-species differences don't manifest in contrasting worldviews and languages.

For example, "Story of Your Life," by Ted Chiang, revolves around a linguist mastering the newly arrived aliens' (very alien) language, and the worldview implicit in it. She succeeds only when

25. Military and intelligence transmissions are encrypted for security reasons. Premium television channels distributed via satellite are encrypted for commercial reasons.

a physicist in the contact team recognizes an alternative perspective to how human physicists usually formulate physical laws. The linguistics/physics conundrums are laid out for the reader in a very striking story.

But from the Department of Special Cases … the SETI discovery or First Contact situation can itself be a trope, the author's way to make us think about ourselves and our culture. Challenges in translation, or the detailed process of learning a new language, are downplayed because the focus is on (a) the message *content* or (b) the unexpected *truths about ourselves* that the existence of aliens forces the human characters to confront.

For example, consider *Planet of the Apes*, by Pierre Boulle. In the novel, far more than in the movies, the ape/human reversal is blatantly allegorical. Major worldview or language differences would only have obscured the point.

### *From which we conclude*

Tropes and translations, alphabets and aliens, worldviews and walruses[26] … this essay has covered a lot of ground.

So in your next alien-featuring story, should you invent an alien language, complete with lexicon, syntax rules, and odd idioms? Should your next time-travel story feature classical Greek? Or should you go the way of the trope?

I can't answer for you. The needs and logic of the story outweigh

---

26. Who guessed I could bring this around to *Through the Looking-Glass*? I could as easily have started there:

> "The time has come," the Walrus said,
> "To talk of many things:
> Of shoes—and ships—and sealing-wax—
> Of cabbages—and kings—
> And why the sea is boiling hot—
> And whether pigs have wings."

That pretty well covers it, I think.

the alien creatures and the temporal displacements. Hopefully this essay has suggested ways for you to approach the questions.

### Back to our story

So why *does* our walrus/lobster friend speak like a Victorian Englishman? Because parody is one more reason to break the rules—

And (check his uniform) he is the very model of a modern major-general.

# *At the Watering Hole*

Resplendent in a spatter-pattern robe, K'reediscranth turned two eyes and four ears heavenward. Two more eyes he pivoted toward the sacred text spread across the lectern against which he leaned. His remaining sense organs scanned randomly across the hushed multitudes far below.

It was his nineteenth Festival of Oneness, and yet the ceremony's majesty threatened to overwhelm him. To the horizon in every direction folk gathered in holy assembly. Myriads more waited beyond his sight. The folk only congregated in such multitudes on the Day of Oneness. Stray tendrils of thought reached him even here atop the holy obelisk.

As High Priest, K'reediscranth had long ago committed the liturgy to memory—the Rite must never vary. With all eyes now focused on The Book of Oneness, he intoned: "In the beginning, the folk were scattered and alone, little more than beasts …"

◆❖◆

Carla Markson dug through the digital debris of a failed career for information that stubbornly refused to reveal itself. Her difficulty was hardly surprising. SETI observations were indexed by coordinates of the celestial sphere and time taken, but she was desperately searching for records of a scarcely remembered pattern she might have seen "about a year ago."

In offices, meeting rooms, and corridors all around, the farewell party had degenerated into a wake. SETI had lost government backing years ago. Now disheartened private patrons had withdrawn support. The funding officially ended at midnight.

"Carla?" an impatient voice called. "Dr. Markson? Everyone's gathering for a final toast in the cafeteria. Please join us."

"In a minute." Images flashed across her computer screen. Two images a second; one hundred twenty per minute. False-color coding now accentuated aspects to which no one had heretofore paid attention. Flash, flash, flash …

"Carla!"

She snapped out of a hypnotized funk. Annoyance became triumph: Had the director not barked at her, the graphic she had so urgently sought would have flickered past her unnoticed.

"Genie, halt," Carla commanded. "Backwards, ten seconds per frame." Data barely sensed returned to the screen. She examined frequency spikes, power densities, and modulation coefficients, all color-coded and textured for pattern recognition.

Two champagne glasses clutched precariously in one hand, Director Harold Flynn opened her office door. "It's *over*, Carla. Please join us."

Plucking the glasses from his grasp, she drained both in the time it took her computer to back up three images. "Genie, forward one frame and stop."

To her boss of the next seventeen minutes Carla said, "Hal, I have a counterproposal. The team should join me here—and bring more of the bubbly." She tapped the frozen graphic with a well-gnawed fingernail.

"I've found them."

◆❖◆

In a pause between chapters, K'reediscranth raised his eyes from the sacred scroll. Gazing over the countless folk gathered for this, his thirtieth Festival of Oneness, he dared to hope: *Maybe* this *year we will succeed.*

Stretching his senses to their utmost, K'reediscranth detected the merest suggestion of an imbalance to the north. The holy assembly, in consonance with his will, shuffled until their aggregated thoughts subsided into a background murmur. He grimaced in concentration until their realignment was seemly.

He resumed reading. "After many an age, our forebears learned to live together as families, to harmonize their minds. The families were fruitful and spread across the face of our world. Individuality remained, but families could meld their thoughts upon need.

"In the fullness of time, families cooperated as clans, then tribes, then nations. Each step enriched the mentation of the folk." Three hands nervously stroked the fabric of his slick ceremonial gown—this festival season, an iridescent rainbow—as he kept his place in the liturgy with a fourth. "In the days of the prophet Holgor'roth, all nations merged into one commonwealth. Their power of thought again grew manyfold."

K'reediscranth bared his senses to all the merged minds roaring their yearnings. The slightest wavering of his concentration, the merest chance interruption of the crowd's careful symmetry around the ceremonial obelisk, and he was lost.

This was the very apex of the Rite.

"The folk strove to become yet greater, and they could not. They sought an even higher unity, but they were already as one. They bore sons and daughters and verblans until the world could feed no more. There seemed no path forward.

"And then Vrg'oq'lan the Prophet revealed a way—nay, *the* way—to the folk."

◆❖◆

The man with whom Carla Markson amiably chatted was short, pudgy, bland—and the King of Sweden. Tomorrow he would present her with the Nobel Prize in Physics. She didn't exactly do physics; for her, the Nobel Foundation had chosen to bend their definitions.

Awarded just five years after her discovery, the prize was an unprecedented recognition. Carla's fellow laureates, likewise mingling with the Swedish aristocracy, were decades older than she.

"Tell me, please," the king said in his charmingly accented English, while nabbing a canapé from a passing waiter, "why you succeeded."

*How to put her insight simply?* Carla searched the ballroom for inspiration. She found it at the crystal-laden, forty-foot-long bar to which liveried servants ceaselessly returned.

She pointed. “Behold the watering hole.”

The original figure of speech, of course, referred to a *water* hole—a literal pool of $H_2O$ in the African veldt—where lions and antelopes and whatnot came to drink. *A woman's reach should exceed her grasp, else what's a meta for?*

“Everyone meets at the watering hole. In the search for extraterrestrial intelligence, SETI, all species were presumed to meet at some celestial watering hole, discoursing over a natural radio frequency.”

“Where is this watering hole?” The king nibbled on wafer-thin toast spread with foie gras, oblivious to the crumbs that dotted his cummerbund.

“There's the problem. People differed on *the* ideal frequency, but they always picked a characteristic frequency of some common molecule or element, like hydrogen. The argument ran that the frequency must be rational and universal, reflecting the analysis that any technological species would make.”

“That's sensible—yet you looked elsewhere.” Her royal companion frowned in concentration as he licked essence of hors d'oeuvre from his fingers. “Why?”

“SETI researchers scanned candidate watering-hole frequencies time and again. They heard nothing.” Reliving her epiphany made Carla beam. “Their strategy contradicted the one fact we had.”

“Which was?” the king asked.

“We consider ourselves a technologically capable intelligent species, yet humans don't broadcast on anyone's proposal for a watering-hole frequency. My colleagues had ignored their only data point.”

◆❖◆

K'reediscranth panted, feeling all his one hundred forty years as High Priest. Next year, he vowed, he would graciously accept an acolyte's assistance to the top of the sacred, and exceedingly tall, obelisk.

He set The Book of Oneness upon the lectern before him. Once his breath returned, K'reediscranth unrolled the scroll and began to read.

◆❖◆

Amid a crew of geniuses, people deferred to Carla. In her humbler moments she questioned their deference. True, she had discovered the civilization, 124.6 light-years distant, to which they would soon depart. Still, was that accomplishment greater than the invention of the FTL drive that made this mission possible?

Admiring herself in her stateroom mirror, snappy in a ship's jumpsuit, Carla decided that yes, just maybe, it was. She exchanged toothy grins with her reflection. Eagerness to meet *her* aliens had saved the FTL project. How much longer, and to what purpose, would star-drive research have continued … if *she* had not saved SETI?

Strident bleating shattered her reverie.

"All hands to departure stations. Repeat: All hands to departure stations. We leave Earth orbit in five minutes." The speaker above her stateroom door added, "We won't postpone our departure even for you, Dr. Markson."

*Nor should they.*

Once she found where to look, archives clearly showed that the aliens had signaled Earth every 302.8 days for as long as SETI had records. Like everyone she knew, Carla felt there was one perfect moment for the embassy ship to arrive. They must appear concurrent with, and as though in immediate reply to, the others' next—and now imminent—transmission.

She hurried to her place of honor on the bridge.

◆❖◆

K'reediscranth poured his hearts and soul into the Assembly, although, or perhaps because, those hearts still syncopated from scaling the obelisk. Beneath a fluorescent robe, sweat matted his thinning fur. One hundred forty-three years in service as High Priest … it was almost beyond precedent.

The so-familiar thoughts of the masses comforted him. He intoned the liturgy, countless minds melding as he read. "For Vrg'oq'lan the Prophet showed us the way. With the world united, more worlds, and their folk, must be brought into union."

"Let it be so," shouted the massed minds of the folk, a roar in his head.

◆❖◆

With the flip of a switch and an instant of vertigo, the blue-and-white-and-brown globe shown on the bridge console transmuted from clearly Earth to clearly … someplace else. The color balance seemed off, consequence of the K-class orange sun. Sensor readings filled the environmental monitor: surface temperature; gravity; rotation; partial pressures of oxygen, nitrogen, and trace gases. All were Earth-like; all clearly not of Earth. Another screen spiked with the characteristic microwave signal that had brought them.

A terrestrial world and a new civilization: grand slam on humanity's first step out of its backyard.

In a flurry of activity the bridge crew put the ship into a synchronous orbit over the single landmass. All eyes then turned to Admiral Peltison. He faced the camera that would beam his words and visage to the planet below. The ETs would not understand his words at first, but surely they would record this momentous occasion. Mission Control had decided early on against beginning the interstellar visit with the universality of an arithmetic lesson. Pomp and circumstance first; language lessons later.

His posture manly but not militant and his voice firm, the commander of the mission from Earth spoke. "Fellow citizens of the galaxy, we come in …"

◆❖◆

"And Vrg'oq'lan spake unto the folk, saying, 'Let a holy assembly be held on the first day of the first month of every year. Let all the world unite on that day. Let the folk reach out with its hearts and mind to what other worlds there may be. Let … let ….' "

Psyches churned as K'reediscranth's concentration wavered. *Damnation*, he thought, striving with purity of spirit to ignore the sudden buzzing in his head. As an acolyte, he'd thought the old priests' complaints of interference absurd. Now he knew better: His mental and sensory acuity had never been stronger, even as this husk of a body failed him.

He swiveled two more eyes to the lectern, the better to concentrate. "Let the sacred assembly now extend itself to the other worlds."

At his priestly command, 1,354,876,202 of the folk, each spaced from his neighbors in multiples of quarter wavelengths, began to emit, in ten-watt pulses, their microwaved yearnings. His guidance, downward-directed from his perch atop the sacred obelisk, steered the resulting multigigawatt maser of prayers from star to star across the glorious reach of the One Heaven.

His voice resonating as the folk sang of unity, K'reediscranth assigned a third ear to the sporadic buzzing. Noise he could handle: He had not been High Priest so long for naught. Without interruption of his chanting, he snapped the focused voice of the folk's desire. As the venerable ones had taught, that quick flick always eliminated the noise.

Decorum restored, K'reediscranth completed the ceremony.

Creeping down the spiral staircase, his crumbling joints protesting each step, K'reediscranth could only wonder: *Why, oh why, was the Assembly of Oneness always held in vain?*

◆❖◆

The fused and lifeless moonlet sailed majestically through the void, the too-orange sunlight glinting unevenly off its tumbling bulk. The only evident hazards to its future serenity came from the large number of like-sized recongealed masses that also orbited the blue-and-white-and-brown world with its single continent.

Little beyond stubborn traces of organic chemicals distinguished this satellite from its many neighbors. The solar wind soon dispersed even those vapors.

And then only scorched paint remained to identify the derelict as the spectacularly ill-named UN embassy starship *Galactic Unity*.

# *RSVP*

Mission Commander Green waited, ill at ease, beneath the hot glare of the media lights. Life was simpler on the bridge of his starship. Nervously, he shifted his weight from one leg to the other. His palms were cold and clammy.

Flashing cameras heralded the entry of the committee into the hearing room. Anxious spectators fell silent. The lawmakers took their places, arranged files, set out their portable computers, poured glasses of water. Anticipating the crack of the gavel, Green took the center chair at the witness table, his crew seating themselves behind him.

He recited his oath. Cameras flashed anew as he gave his prepared statement. Using videos from the recent mission, he recounted the familiar events. The hearing would explore the details, but the evidence seemed incontrovertible.

Yet despite the mountain of data that informed their recommendation, he felt he had somehow failed.

The chairperson began the questioning. "Commander, did your mission follow the agreed-upon protocols?"

"Yes, Ma'am."

"More than once?"

"All the protocols. Repeatedly, and in various sections of the planet." He had been *so* sure of a different outcome. "Time and again, for years."

"Including overflights and interviews with random natives?"

"Yes," Green agreed. "Those were among our procedures. We also flew by and alongside their primitive conveyances."

After keying a note, the chairperson changed topics. "Commander, tell us about the natives."

“Ma’am, their settlements are widely distributed. Their largest population centers almost reach the size of small cities. A few regions exhibit capabilities analogous to an early industrial stage of development.”

Query followed query as legislators took their turns. He had thought himself emotionally prepared for reliving the experience. He’d been wrong.

Memories inundated him. The primitive craft that enigmatically ignored or turned to flee from the mission’s aerospace planes. Aborigines who crashed their vehicles in irrational panic. The natives who lapsed into hysteria or unconsciousness at the mere sight of him. The end-of-mission recourse to improvisation: buzzing population centers. Hovering above cultivated areas scribing elaborate patterns with a pressor beam.

“Commander?”

Behind him, the audience murmured expectantly. What had he missed? “Ma’am?”

She rapped the table until the murmuring stopped. “Commander, how do you interpret these observations?”

Green sensed the gaze of his crew on his back. The scrutiny from the front of the chamber required no such imagination.

The natives had offered no systematic response. They showed little evidence of curiosity. What *could* he conclude? “Despite their use of tools, the most evolved life forms of our nearest interstellar neighbor are not yet intelligent.”

“And?” the chairperson prompted.

Green peered at an image of several aborigines, keenly aware that he was deciding their fate. They were bipedal, but there any resemblance to people ended. The aliens were cloth-covered and hair-topped. Their hands bore too many fingers.

His black, insectile eyes opened wide for emphasis, the mission commander delivered his conclusion.

“I see no reason to delay our colonization.”

# *Unplanned-for Flying Object*

The extraterrestrial object soared silently around the curve of the expressway, spewing intense radiant energy. Powers uncontrollable by mere man kept the object hovering just above the horizon.

As a wave of red taillights flashed toward him, Don Shaw stepped firmly onto his brakes. The crunch of metal and tinkle of headlamps somewhere to the rear announced the presence of a driver with reflexes slower than his own.

In the intensely bright light, buildings and lampposts cast long, knife-edged shadows onto the road. Reflections from the glowing entity glinted off the chrome, paint, and glass of the river of cars before him. Heat from the distant orb struck at him through his windshield. An unseen convection current caught a weathered sheet of newspaper, lifting it over Don's car.

He glanced into the pickup truck stuck beside him. The yahoo behind the wheel squinted, gape-jawed, into the glare. The bill of his backwards baseball cap threw its shadow uselessly behind him. On his other side, a well-coiffed, well-dressed businesswoman tried in vain to ward off the light with a well-manicured hand.

Don pondered the other-worldly object that floated so effortlessly before him. Some of the finest minds in the world had studied it closely. Indirect evidence suggested, he'd read, that the visitor was fusion powered. Federal authorities had tracked its course and extrapolated its appearance with great accuracy, even forecasting its arrival through the local media.

Their warnings had fallen upon deaf ears.

Trapped between exits, there was nowhere to go now but forward. Nowhere to head but straight toward the luminous apparition. He waited for those ahead of him to reach the same decision. They were slow to do so: in ten minutes, he crept forward by less than a mile.

Finally, Don could bear his entrapment no longer. He rolled down his window and stuck his head through the opening. Harmful actinic rays from the distant object struck at his unprotected face.

"Idiots!" he shrieked. "Imbeciles! Buy yourselves a pair of shades.

"The sun is going to keep rising *every* morning."

---

Okay, I admit it: "UFO" isn't about First Contact *or* aliens. It's not even science fiction. The story works, if it did, because you thought it was about one of those things.

Regardless, if you've ever had a west-to-east morning commute, I'll bet that "UFO" spoke to you.

# And If We Are Alone? (Being All We Can Be)

To write science fiction is to be in the what-if business. I'll ask, at least rhetorically: what if we humans truly *are* alone in the universe? (Or so distant from any other intelligent beings that we can never make contact with each other. In practical terms, the effect is the same.)

If so, humanity's pale blue dot is all the more precious.[1]

Still, it would be nice to have other perspectives, other minds, with whom to interact. SF has many stories of earthly species, including dolphins and various primates, uplifted to intelligence. Or maybe, through the power of bioengineering—our own or the blind workings of evolution—the range of what we consider human can be expanded. Or perhaps our figurative children, minds resident in silicon chips, or magnetized spots, or the entangled states of qubits, will arise to keep us mere Mark I humans company. (Quantum bits, aka qubits, are fundamental components of quantum computers. We'll touch on quantum things later in the book.)

The three short stories that follow deal with earthly beings we might someday confront. "Earthly" doesn't require them to be *anything* like us ....

1. The "pale blue dot" was astronomer Carl Sagan's term for an image of Earth captured by *Voyager 1* from a distance of six billion kilometers. At the distance, Earth was reduced in the image to a single pixel.

## *Blessed are the Bleak*

The meek did not inherit the Earth.

Neither did the just, the charitable, the brave, nor even—as I might have expected, considering—the smartest.

I admit I am none of those things. It wouldn't have mattered. Anyone who cares to eat, to have shelter against the elements, cannot survive on the scant resources sufficient unto a Virt.

And that is why, knees knocking and heart pounding, I come to stand on the brink of an abyss.

Preparing to jump.

◆❖◆

This isn't the life I signed up for.

I got computers. I *lived* computers. My house and appliances more or less ran themselves. My car drove itself, navigated itself, parked itself. I spent my commute on the net, and my workday—then surfed, IMed, and tweeted yet more once I got back home. I owned an 88-inch holovision, often found my options wanting, and did a fair amount of virtual-reality gaming.

I should have seen this coming.

Not that many years ago, I was just another tiny cog in a really huge machine. Sure, the world was changing, faster and faster every year. I'd have admitted that. But some things were constant. Death. Taxes. More and more money expended in the futile effort to put off death. Healthcare was a quarter of the economy and still rising. What position could be more secure than something implementing cost controls in the healthcare industry? None, surely.

Except the situation I fell into: a civil-service slot when healthcare got nationalized. At Universal Care. The big UC.

Just don't get caught by management pronouncing that "yuck."

◆✣◆

I brace myself against the gusts that threaten to send me, unprepared, over the cliff. Whipped by the frigid wind, my cape flutters behind me.

It's not truly cold, I know. Somehow it feels cold. The chill breath of mortality.

Seabirds circle overhead, piping mournfully. The air has a salt tang and a bit of fishy smell through the nose plugs of my headset. The gear wouldn't be of much use if the pickups blocked sensation. Anything I fail to see or hear or smell or feel, the Virts won't …

Is *sense* the right word? Thankfully, I don't know. Trying not to think about it, I'm unavoidably aware of a zillion tiny prickles from the microelectrodes of my snug-fitting helmet's neural interface. My scalp is suddenly crawling …

Does *that* show up on the recording? Maybe, like my fear, it's part of the experience.

Far below, the ocean roars. The surf crashes. Craning my neck, I peer down. My goggles superimpose an irregular outline over the crashing waves. The thin red line marks the supposedly safe landing zone: between boulders and where the water is deep enough—barring an untimely trough—that I might not pound myself like a tent stake into the sea-bottom muck.

Another day, another dollar.

◆✣◆

Maybe you've seen me online. In more than one virtual world, I was the Caped Avenger: fast with a sword, faster with a healing spell. With one tool or the other, I succored the weak and protected the blameless. I undertook noble quests. I meted out justice. In hindsight, I was using VR to balance a cosmic scale.

Because with each passing day my job, more than anything else, involved implementing new ways to say "no."

And as I sought, in vain, for relief, some of the newest players got good. Damned good. Impossibly good. Turns out they were more virtual than me. Uploads, they called themselves: ascended to a better place. Upchucks, one of my gamer friends called them: some things got lost in the translation.

But one thing I have to admit about Virts—they were masters of the games. You could say they had been reborn to them.

I should have seen this coming.

◆❖◆

Don't blame me for the waiting lines for care. Or for disallowing treatments as medically unnecessary, or experimental, or off-label, or "too risky" given the age of the patient. Or for … whatever policy stymied you. I only maintained the software that made it easy to do those things. And when "queue management" failed to reconcile an aging population with a shrinking work force, it was suggested to me that a bit of electronic misfiling might help. Most people would wait for a while before following up. Or forget. Or give up. Or die.

You *can* blame me for being an enabler.

And so later and later into the evenings the Caped Avenger did his best to do virtual good. At least enough good so that afterwards I might sleep.

Only the Virts, more of them playing by the day, made doing good harder and harder. Virts were *mean*. I didn't understand their problem. Would they have been happier checked out altogether?

I should have seen this coming.

◆❖◆

The more people become Virts, the faster the economy deflates. Implodes. Few people actually make anything anymore. Mostly they shuffle information in and among computers.

I'm damned good at that.

But hardly as fast as someone already *in* the computer. And when your room-and-board budget only has to cover a bit of silicon and a trickle of electricity, you can underbid anyone who is still flesh and blood.

And the little work that still requires hands? Every new Virt means a bit less need for farmers, or construction workers, or—you name the trade. More than enough people who *know* those tasks compete for ever fewer jobs. Dilettantes need never apply.

Which leaves …?

I turn my head, trying to psych out the motion of the waves, the probable undertow, the hazards of flotsam and jetsam and sharks. The red outline marking my target shifts as I move my head, but as for any other dangers my headset gives no clues.

You Virts *will* have your fun.

◆❖◆

I was without a clue when Reggie hobbled into my office that day and plunked his e-reader on my desk. It wasn't only the hesitant gait: he looked prematurely old. He had some rare disorder, one of the orphan diseases, a condition he didn't discuss. Maybe if I had understood that his prognosis was all downhill. Or that the only possible treatment was personalized gene therapy, very expensive, for which the UC waiting line was out the wazoo. Maybe.

But he was just … Reggie.

"Hi, Malcolm," Reggie said.

"Morning," I replied. "What's up?"

"Good choice of words," Reggie said. "Uploads fix everything."

The odd statement brought to mind film cameras. They still existed, not that it was easy to find the film. The people who used them—the people who inexplicably thought their antiques were the norm, to be fussily distinguished from *digital* cameras—were our most expensive clients. Fixing was the magical, near-medieval process of rendering permanent an image from film. Chiseling the image, if not quite into stone, in silver halide.

Synapses are strange and wondrous things. I didn't get uploading them.

"Have a seat," I told Reggie. "I'll bite. What, exactly, do uploads fix?"

"Budgets," he said. He swiveled a guest chair and settled, legs casually astraddle, elbows folded across the thinly upholstered back. Smug as hell about *something*. "Fixes UC's budget forever."

Reggie was a political junkie, an amateur policy wonk, who refused to get that not everyone was. But Reggie was also our office's lead sysadmin. If you've never programmed for a living, that title won't mean much to you. Think of him as the office manager, the boss's secretary, or that guy in Accounting who has to sign off on your expense reports. Someone on whose bad side it's most unpleasant to be.

"Fixes how?" I asked.

He tapped something on his e-reader, and a headline popped up: *AG Amicus Curiae in Radowitz.*

"Uh-huh," I said. I understood one word of that headline. The preposition.

"Friend of the court. The attorney-general is testifying today at the Supreme Court."

Okay, that left only one word to explain. "Radowitz?"

"Bazillionaire hedge-fund manager. Got himself a terminal case of H5N1. Put his money into a trust before he went. The son is the trust's beneficiary and he's suing to take control. Technically, the case is Radowitz v. Radowitz Trust."

H5N1: avian flu. An unpleasant way to go, but all too common and surely not Reggie's point. I wondered how much money had to be at issue before an estate dispute percolated up to the Supreme Court.

"Uh-huh," I said again.

Reggie laughed. "Radowitz Sr. became a Virt."

The fog began to lift. "So Senior, uploaded, still controls the trust."

Reggie nodded. "Make the trust pay out, and the attorney managing the trust for Senior loses a very cushy job."

A four-wheeled dolly, one wheel clattering, wobbled past my open door. More office furniture for the ever-growing empire that was Universal Care.

"And uploading fixes something," I said dubiously. For Junior or Senior? I had no idea.

◆❖◆

As I shiver in the wind and cold spray—and a bit, too, from the rush of oppressive memories—two quick clicks sound in my headset earphones. "We're not paying to watch you stand."

Virts, among their many deficiencies, have no patience. But maybe if my brain had been disassembled synapse by synapse for readout, I'd be testy, too.

What the hell does it matter? However long I take to work up my nerve, the delay, like those two clicks, like the nag, will be edited out. I say, "I'm waiting for the gusts to let up a bit."

Life would be much simpler if I wasn't naturally healthy.

◆❖◆

If Radowitz Sr. was still alive, his bazillions remained his. The money would stay in his trust, and Junior would wait forever for the wealth over which he salivated. And the feds would not get their cut.

I quoted my mom, quoting some senator from her youth. "A billion here, and a billion there, and soon you're talking real money. No wonder the AG is there. Estate tax."

Reggie smiled crookedly. "What's a billion compared to health-care? Peanuts. We're talking precedent here. Watch and learn."

He had CNN streaming in a window on his e-reader, and shushed my meandering speculations when the AG appeared on the steps of the Supreme Court to give her statement. She said, "Is a person less a person for wearing glasses? Having a hip replaced? How about one of our fine young men and women in uniform, sent home without a limb to be fitted for a prosthetic? The measure of a man or woman is not what fraction of their body is 'natural.' "

Lawyers will argue anything, I thought.

I tried again. "Why are we—"

"Just watch," Reggie said.

We watched. I fidgeted with a stapler as I tried to parse Reggie's interests. (Don't ask why I had a stapler, but no desk could be without one.) If the Supremes ruled that a Virt was alive …

"Are you saying uploading will become … a *medical* procedure?"

Reggie nodded. "Readout costs nothing compared to bypass surgery, or hip replacements, or … choose your poison."

Poison: that's what, at Yuck, we called chemo. And radiation. And a nice, personalized retrovirus cocktail. That last one was on Reggie's mind, and I had no idea.

I knew no Virts personally, but the VR games had become overrun with them. Shadows of former people. Mean shadows.

Bored shadows?

"Okay," I said, "readout is cheap. So what? Who would choose uploading over, say, a hip replacement? You upload when you *have* to."

"Choose?" Reggie smirked. "Yuck doesn't let people *choose* their course of treatment. You accept the recommended treatment, or you're on your own. And the frosting on the cake: after uploading, a Virt never incurs another medical cost."

"Why are you so into this?" I finally thought to ask.

"Did I not mention? The government siding with Senior wasn't the AG's idea; it was at the UC's recommendation. Nor was it the brainstorm of some Yuck High-Muckety-Muck, either.

"Not to be modest"—Reggie buffed the nails of one hand on his shirt—"*I* suggested it."

◆❖◆

Maybe Virts *are* alive. So the Supremes have ruled, and so they gave us our first real shove down the Teflon slope.

Suppose the Supremes are right. Then Normals and Virts are at war! And for a mess of retroviruses—rather than Upload himself—Reggie sold out his own kind.

Seniors voted more than any other demographic. That led to untouchable entitlements and almost to national bankruptcy—until Radowitz v. Radowitz Trust. Now Virts vote, too, and *they* can't imagine the government spending too much on VR. Their "homes" are many-times fault-tolerant and widely distributed.

And still a bargain compared to medical care.

As Universal Care morphed into Universal Computing, I agonized. The Virts were taking over—everything. Nothing short of Armageddon could physically stop the process.

Could I destroy enough copies to make a difference? No. Could I plant a sufficiently effective virus or a worm? I looked and looked and never found a way.

Could I do *anything*? The conundrum kept me sleepless every night, got me moving every morning—

Until the day I found out I had missed my chance.

Yuck had laid me off. Henceforth, for a pittance, a Virt would be doing my job.

◆❖◆

The wind died. Trying to ignore the omen, I leap.

Seventy feet. The math is simple: a hair over two seconds.

The waves rise up to swat me. Math be damned, they take their time. Long enough for my life to flash before my eyes. Long enough to think:

—The meek have not inherited the Earth. The mean did.

—Given my deeds as a cog within Yuck, I am little better than they.

—Maybe no better.

—That my cape is a stupid affectation.

—How ironic it is that Reggie, after the Faustian bargain by which he got proper treatment for his illness, was flattened by a wayward bus. Smashed, even, beyond uploading.

—As, quite possibly, *I* am about to be flattened.

And finally, as the surface rushes toward me, to think: *this* is my life henceforth. One spectacular act of daring—of choreographed insanity—after another, until my luck runs out. Every merely reckless and painful entertainment has been recorded already. I've looked.

And so, here I am.

Waves lunge at me. Massive, jagged rocks stare hungrily at me through the spray. The air buffets me and the red outline of my target drifts.

*This* is my life. Unless …

I had never found a way to plant a suitable virus. But what about a meme?

I twist and stretch my body to alter my wind resistance. To swerve *outside* the boundary. Thinking dark—honest—thoughts about what we all, Virts and Normals alike, have become.

There is an instant of excruciating pain—

◆❖◆

My eyes fly open.

I'm not sure where, or even who, I am. I'm dead, surely. But faster than I can marvel—there *is* a heaven—the fog begins to lift.

I can see—sort of. And hear—sort of. And I know who I am. *What* I am.

Not Malcolm Jenkins.

And yet a trace of Malcolm persists: a ghost in the machine. Enough to be horrified by the giddy digitalized persona of—among how many others?—one Elizabeth Tyler Andrews.

Who is excitedly queuing up the madly popular feelie for another playback …

# *Chance of Storms*

My cabin, its lapped wood siding faded to a pale gray, was the only human structure in sight. I was communing with nature, aka staring blankly into space from the front porch, my chair tipped back, my feet propped on the rickety railing, when motion caught my eye.

A plume of dust made its way slowly along a path that was more hinted-at than beaten. In time, a vehicle became visible. Watching the unexpected car bounce and sway, I wished upon it blown shocks, flat tires, and a snapped axle. Such a minor mishap would have been a blessing—for whoever was in the car.

No such luck.

The road became ever more rutted as it approached the cabin. My uninvited guest could not have been going more than twenty when he passed the *Do Not Trespass* sign. He ignored it.

I sighed.

The late-model Toyota, arguably metallic blue, but I couldn't be certain due to the coating of dust, shuddered to a stop a good hundred feet from the cabin. There was an inaudible consultation among the three people inside, then the driver climbed out. He wore an expression that was simultaneously aggressive and anxious, a look with which I had become all too familiar: a reporter. "Mind if I come on up?" Mistaking silence for assent, he advanced to the foot of the porch.

He turned slowly, taking in the broad expanse of prairie that surrounded my humble home. "Nice view. I can see why you would like it out here."

Who said I liked it? I continued my silence. Sometimes rudeness worked.

"I'm Harry Weidner." He named the national newspaper he

claimed to represent. I waved him off when he set foot on the lowest step to show me a press ID. He put away his wallet, then ostentatiously patted his shirt pocket. “I should tell you that I’m wearing a wire. The signal is broadcast to my peers in the car. Recorded there, too.” He gave an equally ostentatious shrug, and added, “I’m sure you’ll understand.”

It was my turn to shrug. “I don’t suppose you would consider getting back in your car and just leaving?”

“I don’t think so.” He flashed the reporter’s standard this-won’t-hurt-a-bit smile. “This is a story that should be told. I’ve done a lot of research about you, you know.”

It would do no good, but I felt obligated to make a disclaimer. I gestured at my, shall we say modest, surroundings. “Are your readers crying out for news of obscure failed writers?”

Weidner licked his lips. “Only the ones with paranormal mental powers.”

A storm had come up suddenly. Clouds were forming to the west; distant thunder rumbled. “Since you have researched me, what do you think my story is?”

He looked around for a seat, since I was already occupying the only chair on the porch. He settled for a tree stump back in the yard. “It seems clear that you have paranormal abilities—we’ll come back to which ones. As those talents matured, you’ve moved farther and farther into the country. Your former neighbors all attest to strange and unexplained events, especially just prior to each of your moves. *Destructive* events.”

Only the surviving neighbors attested, I assumed … unless Weidner had his own unusual skills. That thought brought on an unwelcome rush of memories. When I returned to the present, my uninvited guest was still droning on.

“… Recurrent incidents of mind reading. That would be consistent with your repeated moves, each time to a smaller and more rural community. To get away from the mental clamor of others,” he ended helpfully, as if I were too stupid to get the point.

Did telling obviously frightened people not to be scared constitute mind reading? Evidently. That was back in the days I deluded myself I had any control.

My guilt-in-advance kicked in. "May I offer you some water?" I retrieved a pitcher and glasses from inside the cabin. "Mr. Weidner, no human sense works that way. Our brain lets us tune out unwanted background noises, ignore unimportant visual detail, stop noticing unpleasant smells after a while. Why would a telepathic sense be any less practical?"

"So not a denial of telepathy—just a disclaimer that it would be a reason to move." The reporter stuffed his hands into his pants pockets. "What about your telekinesis?"

My cabin lacks electricity; the water was tepid. I took a long drink anyway. "If I could move things mentally, don't you think I would have brought us some ice cubes?"

He rattled off several incidents from my past, odd occurrences of various kinds. Would I admit to having clairvoyance? Prescience? One might say these speculations were closer to the mark, but I didn't bother. Any foreknowledge of doom I experienced was simply pattern recognition. My vague disclaimers continued to be, in turn, disbelieved.

I don't know why Weidner finally left. Maybe he finally accepted that I wasn't going to tell him anything. Maybe the approaching storm made him decide it was time to head back to town. Whatever the reason, his nattering finally wound down. He strolled back to the car. He pivoted to make one last attempt. "Fine, then—I'm all wrong. My research is all wrong. So what *is* your story?"

My only answer was a glower; he eventually got into the car. The vehicle turned in a tight circle to head back the way it had come. It had gone perhaps two hundred feet when a rut caught a front tire. The Toyota swerved into a roadside boulder and halted with a loud crunch. Airbags briefly filled the windows. Steam curled upward from the cracked radiator.

Although three people climbed out under their own power, the familiar pangs of guilt returned. With fear, or at least a too-late sense of doubt, now plain on his face, Weidner watched me step down from the porch. He slumped with relief as I turned away from the wreck.

Fat raindrops, harbingers of what would surely be a major downpour, flung themselves from a suddenly slate-gray sky. Coyotes howled in the distance. Thunder boomed, much closer now, as my uninvited callers began the slog to the main road.

Walking briskly in the opposite direction, I explained to no one in particular. "My story is … I have contagious bad luck."

# A Time for Heroes

Travis Logan was a reader, a movie buff, a TV addict, a gamer. In short, he was a lover of stories—but not just any story held his interest. He reveled in grand sagas, epic explorations, daunting quests, and perilous adventures. The more larger than life, the more fascinating the story—

And the more Travis's own mundane existence suffered by comparison.

In the stories he loved, dire circumstances forged ordinary people into heroes. The situation might be cruel or capricious, the work of an implacable enemy or of vast, impersonal forces. A debilitating illness. A car crash. Wartime horrors. Zombie plagues. A lover in peril.

It could be anything. But it had to be *something*.

Alas, for as far back as Travis could remember, he had had no such formative moment. There was no deep, meaningful lesson to define him. In all his humdrum life, there had been no opportunity to seize, no peril to overcome, no noble sacrifice to make, no grand challenge to which he might valiantly have risen. The one incident involving Travis that might fairly be called seminal was, quite literally, seminal, and his parents' experience rather than his own.

No, Travis had never had a life-altering moment. Perhaps that was why his life was so commonplace and seemingly of no account.

The thing about life-altering moments is, they do not often announce themselves ....

◆❖◆

Mortars roared, and the ground shook beneath Travis. Rock chips flew from the boulders behind which he crouched, as bullets screamed past. Tendrils of mist writhed overhead, lurid with crisscrossing laser-sight beams. His hands hurt from clutching a rifle so tightly.

This was the best game ever.

*Here* was far more real than the sensory-immersion tank in which Travis knew himself to truly be. His senses were alive, alert to the smallest detail. Hands down, this was the best virtual-reality environment he had ever experienced.

He looked around. His uniform, even the armored vest, was all tans and greens: forest camouflage. Every blade of grass was distinct—except, he noted, when looking out the corner of his eye. The clouds far overhead were like that, too, and the texture of the boulders: details fading on the periphery to conserve computing power. Done right, the optimization shouldn't have been visible. It needed tweaking.

But he wasn't here to critique the graphics, or sound effects so authentic that he felt them in his bones, or the imaginary pebble so convincingly digging into his butt. His job was putting the game through its paces. Playing.

Only this hardly felt like play. A mortar round landed, the closest yet, shaking the ground. Dirt, twigs, and pebbles rained down on him. Soon the imaginary mortar would imaginarily find its target. *It's only a game*, Travis told himself. *Take your time. Scope things out.*

Boulders ringed him. He crept toward a gap between the rocks to peek out.

Travis never saw the sniper who killed him.

◆❖◆

Bullets whizzed overhead, as many as before. But unlike before, he was going to *move*. He had a mission to accomplish. He was Zorro with an M-16.

Travis ripped open vest pocket flaps, the Velcro zzps inaudible amid the gunfire. Eventually he found a little mirror on a short handle, a gadget like a dentist might use. That was new, and he smiled. You had to love online gaming. Unseen developers had added the mirror to his gear.

Carefully maneuvering the mirror between and above various boulders, he scoped out his surroundings and even spotted a few of the opposition.

His watch was busted, by a bit of ricocheting rock perhaps. He estimated the time between mortar rounds by his pulse. Call it eight seconds.

When the next round hit, Travis was on his feet while the ground still shivered. Under cover of the smoke and still-falling dirt, he dashed for the nearby woods.

A camouflaged enemy, smirking, emerged from behind a tree and shot him.

◆❖◆

Spraying the trees with automatic-weapons fire as he ran, Travis made it safely into the woods. The enemy side kept learning—but he learned faster.

He caught his breath, standing over an enemy soldier dramatically dying. Too Hollywood, but not bad acting for an artificial intelligence. In gamer-speak, a bot. A game-playing, robotic character. That's what all his opponents were: software.

Software embodied in graphics so—graphic—that Travis had to look away. He felt queasy. He told himself he was being ridiculous. The throbbing wounds were no more real than the characters he was here to train.

He was a pro gamer. Lots of companies brought him in-house for a day or a week before releasing a new virt game, to run a final checkout. To test, not to put too fancy a term on it, untainted by knowledge or preconceptions about the game.

But this was much more than testing, as vague as his host had been.

Scripted bots were *so* last century. They did what the developers instructed and no more. Watch a scripted bot for a few games and you could kill it every time. Where was the fun in that? Or the repeat customers to keep renting time in the very expensive VR tanks?

So most virts added randomness to their bots. The sniper who stood behind a tree today would almost certainly hide standing behind a different tree the next time he played. The variability made play more challenging—until, in no more than a few iterations, the gamer learned to study all the trunks at the forest's edge.

The bot ostentatiously wheezing at Travis's feet had lain prone behind a fallen, rotting log. Better, but still hiding behind a tree trunk. Travis looked around. This sniper could as easily lurk behind that low stacked-stone fence, or down in that meandering streambed, or—well, that was the point. That the gamer not be able to anticipate. A *trained* AI—that's what would make this virt a moneymaker. And Travis was here to give the training.

"You're too easy to spot among the trees," Travis told the enemy. "The log isn't different enough. Mix it up more. Remember that."

With one last flamboyant shudder, the bot died. Its last gasp did not sound much like, "Thank you."

And Travis, in the seconds before the dead bot dissolved into a puff of greasy smoke, didn't feel much like Zorro.

◆❖◆

*I'm getting the hang of this virt*, Travis decided. He was deep in the woods, well on his way to his objective, a bevy of dead enemies marking his passage. Shot, grenaded, knifed …

Of course *he* had died many times, too, learning to get this far.

He was running low on ammo, and he went through the pockets of the soldier dying at his feet. Travis found three ammo clips, half a dozen grenades, and a pair of binocs.

Stealing from the dying hardly felt heroic. Did the Scarlet Pimpernel ever do such a thing? Odysseus? Batman? *It's not real*, Travis reminded himself. And that he would have waited for it to die, if the game software didn't remove the body so soon after. *It's not real.*

In some still, small recess of Travis's mind, Batman rebuked, "Neither am I."

◆❖◆

Travis lay flat on the rough ground, binocs in hand, peering out from under a dense thicket. An enemy outpost blocked the obvious routes forward. He could try to sneak past by dark of night, but they would expect that. And the sentry he had garroted had had night-vision equipment. Others would, too.

Dark of night: When would that be? He couldn't remember the apparent time of day he had begun playing. For that matter, he didn't remember whether it was the same every game.

Hmm. How many times *had* he started over? Curious. Travis couldn't say. Still, he couldn't have been in the virt all that long. He'd had only a modest lunch and he wasn't hungry.

He glanced at his virtual wrist and found his virtual watch shattered again. He shrugged. When next he died and came back, he would take better care of the chronometer.

For now, not in a mood to wait or die, he studied the terrain. *There*, he decided. A good place for a diversionary explosion or two. And there, an even better spot for waylaying the enemies who must come to investigate.

He crept forward to lay an ambush.

◆❖◆

For all Travis's progress, the objective still lay far ahead. It was all about the mission now. When, inevitably, he "died" from time to time, he returned through enemy lines with ruthless efficiency. Deaths were the price of advancement.

He disabled tripwires, bypassed sensors, and dispatched enemies, all with casual ease. A trail of bodies littered his wake. The bots learned, but not as quickly as he. If they could, this would be one *hell* of a game. But they were so much cannon fodder.

Soon, he would reach the target. He would destroy it. *Just like in* Lord of the Rings, Travis thought. *No matter the odds, no matter who tries to stop me, I shall prevail.*

Only why did he feel more like some cyborg terminator than like Frodo?

◆❖◆

A bot lay at Travis's feet. Sure, they were wily—much more so than when he had begun playing. He had learned, too. Natural intelligence still ran rings around the artificial kind.

Apart from some bruises and the rivulet of blood that ran from his torn cheek and down his neck, the prisoner looked familiar. Why wouldn't he? The game software would have only so many faces to put on bots. Travis must have seen this face again and again.

And killed it again and again.

"You're going to talk," Travis said.

"No, I'm not," the prisoner said. Hands tied behind his back, he struggled to sit up.

When he had almost succeeded, Travis planted a boot on his captive's shoulder and knocked him down. "I need to know the route ahead. You're going to tell me. If I come across anything that looks even the slightest bit different than what you describe"—Travis unsheathed his commando knife, its double-edged blade a wicked seven inches long—"I'll be back."

"Who cares?" the prisoner snapped. And then—

In a puff of smoke, he vanished.

◆❖◆

Travis loped upslope, gliding between the trees, skirting an enemy outpost on the valley floor. His mind churned.

Bots disappeared all the time—after you killed them. Removing them was merely fancy graphics and practicality. Dead bodies everywhere would clutter the playing area and waste computing cycles. But for a live bot to vanish? That made no sense.

Unless that character wasn't a bot.

If it wasn't a bot, that made it … a player. A gamer. A *person* Travis had threatened to torture! And if one enemy was a real gamer, more would be.

Travis skidded to a halt, chest heaving, within a small stand of trees. How many gamers had he casually slaughtered? How many repeatedly?

Sure, technically he only killed software. Knowing some of his victims had people behind them felt different. Felt *wrong*.

It was time for a break. First thing, he'd apologize to the gamer, assuming he could track down who that had been. There must be log files on the game server. Only then did Travis realize—

He had no idea how to get out of the virt.

◆❖◆

He sat on a fallen, mossy log, on the edge of a small clearing. Faint noises drifted his way: enemies hunting for him. He hoped they could tell him the way out.

He knew one way that did not work: dying. He had died in the game …

He could not remember how many times. Or how long he had been playing. He remembered planning to check his watch when he reentered the game. He hadn't. He glanced at his wrist.

The watch was broken. Maybe it always started that way.

He tuned out the hunt, struggling to concentrate. Exiting the virt should be a simple matter of opening his eyes. Raising the lid of the VR tank.

But his eyes *were* open. Here, anyway. He closed them and extended his arms into the dark. Groping in front of him, then overhead, then all around him, he felt … nothing. Beneath there was only the rough bark of the log and the cool dampness of moss. No hint of a VR immersion tank.

A part of Travis still needed to struggle onward. How else could he reach the objective? How else could he complete … his … mission?

He twitched. What, exactly, was the mission? He had no idea. He wondered if he ever had. The game, like this computer-generated valley, might go on without end.

Leaving should be the reverse of entering. He had played many VR games, in many such tanks. Only he had no recollection of getting *into* a tank this time. Odd.

It was surprisingly difficult to retrieve memories from right before the virt. They seemed distant, as though he had been playing for a very long time. His last pre-game recollection was, was, was …

A computer lab. VR tanks. A tall, moon-faced man, standing. It all felt less real than the virtual log on which Travis "sat." He frowned in concentration. That man. Cav something. Cavender? Cavendish?

Cavanaugh! That was the man's name. Alan Cavanaugh. He was Travis's host and the company's creative director. Cavanaugh held a helmet, dome down, like a hardhat lined with tiny silvery Lego blocks. Sensors, of course. Lots and lots of sensors. EEG on steroids.

Cavanaugh said, "I'll explain the game in a minute. For the best experience, we'll configure the tank especially for you." He offered Travis the helmet. "Put this on. This will read out your ideal settings."

Travis remembered thinking that a personally calibrated VR tank would be awesome. He remembered taking the helmet out of Cavanaugh's hands and setting it on his head—

And appearing, dressed in camo, pinned down by enemy fire, amid a jumble of boulders.

◆❖◆

Travis sat tailor-style on the ground in the center of a wooded glade. His hands rested on his lap, palms up and fingers interlaced. His rifle, well beyond his reach, its ammo clip removed, leaned against a jagged stump. "Let me out," he said to the air.

No one answered. Nothing happened.

He waited, looking harmless.

Enemy troops eventually rushed into the clearing. Most were expressionless: clearly bots.

One shot Travis faster than he could ask the way out.

◆❖◆

Out of the boulder field and across the meadow, into the woods and deep into the valley, Travis had once again left bodies in his tracks. The mayhem had nothing to do with a "mission" and everything to do with his need to hide. He had to disappear, to make the enemy spread out hunting for him.

Now he hugged a hollow in the ground, beneath a thatch of torn tall grass. A scout approached, circling inconsequential bumps and dips in the terrain with the telltale algorithmic fussiness of a bot. Travis let it pass. Eventually a second appeared: another bot. Travis allowed it, too, to go by unmolested. And then—

Nuances of motion and stance said *this* was a gamer. When the prey turned his back, Travis burst out of hiding. The enemy scout stiffened as a rifle barrel prodded him in the back.

But this *wasn't* an enemy, only a link to the outside world. The enemy was whoever kept Travis inside the game.

"How do I—?" Travis began to ask. Only no speech, no sound at all, came out of his mouth. The programmers had been busy again.

He was too surprised even to notice what killed him this time.

◆❖◆

Amid a too familiar jumble of rocks, bullets zinging overhead, mortars blasting, ground trembling, Travis considered. There was much that he remembered only imperfectly. The tastes of foods. The colors of a sunset. The clothes he had worn to the "test" session. The names of the game and gaming company were completely gone.

And yet.

He remembered enough to know he was no mere game character. Who would fill a bot with the names of kindergarten classmates, and the capitals of the fifty states, and the smell of freshly baked bread? His memories, however incomplete, *were* his.

Scripted bots. Randomized bots. Artificially intelligent bots. He was none of those things. What came next? No question: Naturally intelligent bots. Uploads.

*Me.*

Target for an ever-growing legion of players who would seek to kill him. Again, and again, and again.

In a flash of light and sound, a mortar round found him.

◆❖◆

Rocks. Bullets. The roar of mortar rounds. Deep in thought, Travis ignored them all.

He had been *copied* into this game.

The real Travis would know the goal of the game. *He* would remember getting into a VR tank, and know how to open it.

Would an uploaded copy know? Apparently not. Or rather, for the copy there was no tank. He was only bits inside a computer, the ghost within the machine. There was no door for him to open.

The mortar rounds crept closer.

For what it was worth, Travis knew when the copy—he—had been made. Personal calibration? Ha!

Was he a bootleg upload? No way. Flesh-and-blood Travis might log onto this game any time, recognize himself, and sue the company for everything it was worth. Why risk it?

No, Real Travis had agreed to this. That Travis didn't remember the discussion only meant Cavanaugh brought it up after the copying process. Real Travis would have jumped at the chance to star in a virt—and happily collected his royalties.

*Because* I *would have*, Travis thought sadly. *And I'd have been wrong.*

Blam!

Clods of dirt rained down. Travis's ears rang. His nose bled. That was *close.* In another round or two, inevitably, the enemy would find the range. In a few scant seconds …

He didn't feel much like Batman, Zorro, or Frodo. He didn't feel inspired by them. He didn't feel like fighting, or even like moving.

Travis remembered wishing life had—even once—dealt him a real challenge. A worthy challenge. A seminal moment. The type of defining moment that forged true heroes.

Real Travis, yielding to vanity or cash, had had *his* seminal moment. In dooming his own upload to purgatory, Real Travis had shown himself unworthy.

*And me*, Travis thought. *What about me? This is a seminal moment.* My *seminal moment. I choose to fight no more.*

*My hero now is Gandhi.*

As the whistle of an incoming mortar shell swelled, Travis wondered how many times, and in how many uninteresting ways, he must die before the game masters ended the slaughter.

# *If This Goes On*

Do you often sit around contemplating what humanity will become—or what will replace us? Likely not. It's taxing enough to wonder (worry?) about What Comes Next. My guess is most people do that. Which brings us to trends ....

Whether they continue or stop, trends are confounding.

Here's a trend: incorporating radio-frequency identification (RFID) chips in more and more consumer goods. Not familiar with RFIDs? For now, think of them as spies in a garment label. The technology will be fully explained (and made scary) in the essay, "Beyond This Point Be RFIDs."

In the short while between that essay and this collection, the Internet has changed dramatically. What makes an RFID most valuable is its ability to report its location. What makes the technology scary is the ease with which millions of RFIDs and *billions* of their readouts are tracked in online archives. Like tweets and Facebook posts, RFID readouts aggregate into vast databases that can be sifted and sorted, used and abused, in ways never intended by the originators of the data. (And whether a database was intended to be Internet-accessible scarcely matters anymore. Ask Anonymous.)

How quickly does the worldwide web change? The web, first available to the public in 1991, began as a source of static content provided by the owners of web servers. Think: scientific papers, news reports, and company catalogs. The second-generation web added dynamic content, most notably user-published material. Think: blogs and social networks. The third generation, rapidly descending upon us, will be the semantic web.

Think of the semantic web as adding logic to search capabilities. Software can begin to exploit the web to draw inferences and plan courses of action. In generations one and two, the web has been people-operated. The richer languages and data models of the semantic web will enable machine-centric operations. Imagine having a personal software agent that searches on your symptoms within medical databases, and then diagnoses your condition(s) by analyzing what it's "read."

I picture Web 3.0 as a step towards an AI-friendly Internet.[1]

That brings us to the second trend for this section: smarter and smarter computer software. How far might that trend go? How humanlike might an artificial intelligence get? Check out "Where Credit Is Due."

Will the trends in this section continue? We shall see.

1. As I discussed in "O Brave New Web" http://www.uiaa.org/illinois/news/blog/index.asp?id=79).

# *Beyond This Point Be RFIDs*

## (First published in September 2007)

"This is a consumer alert," the caller begins. The synthesized voice represents your favorite grocery store. The nut-free cookies you recently bought have been found to contain traces of peanuts.

"This is a consumer alert," the letter opens. Two tires on your car have been recalled. You wonder how the manufacturer found you, because you purchased those tires with cash, three years and two addresses ago.

Ten computers vanish from your place of business. Suspicion falls on you when the automated security system falsely reports that you were in the building that Saturday.

You're traveling in a foreign country. Terrorists snatch a handful of people from a public square—and it is your misfortune to be among the abductees.

### *Enter the RFID*

One technology—radio-frequency identification—unites these scenarios. In a little while, we'll explore how radio-frequency identification, RFID for short, influenced these events. We'll also consider societal and science-fictional implications.

First, let's review the technology.

Before RFIDs, there were barcodes. A barcode represents information—say, a product identification or part number—as a visual pattern imprinted onto the surface of the tagged item. On consumer goods, such as a can of peas, the width and spacing of parallel lines encode the data. Computerized checkout, inventory management, and asset tracking are among the common applications. At its most basic, an RFID tag also encodes and makes available data.

So far, the RFID tag may sound like the familiar barcode. But an RFID tag is much more, because it can announce itself without a human in the loop.

An RFID digitally encodes its information as data within a semiconductor microchip. RFIDs can (though not all do) hold far more data than barcodes. The chip does more than store bits, however. Minimally, it also provides a radio transponder (transmitter/responder). The complete RFID tag pairs a microchip with an antenna. In physical size, the antenna dominates, varying with the intended application. The chip itself can be grain-of-rice sized.

RFIDs come in two basic types: active and passive. An active tag contains its own power source. A passive tag draws its power from the radio signal that polls and awakens it. Passive tags modulate the reflected portion (also called "backscatter") of the received radio signal.

(Some analysts identify a third type, called either semi-passive or semi-active. The hybrid uses an onboard power source to run its integrated circuit—active—but signals by modulating the RF backscatter—passive.)

Active tags transmit at higher power levels than passive tags. All other things equal, active tags can be read reliably from greater distances than passive tags. (How far? That depends, as discussed below.)

Not surprisingly, passive tags cost less. Even in volume, active RFID tags typically cost dollars apiece. In sufficiently high volume, passive RFID tags cost only a few cents each. Still, even passive tags aren't yet so inexpensive that you'd be likely to find one on that can of peas. You are likely to find an RFID tag in, for example, many garments.

Not all RFID tags go by that name. Embedded RFID tags enable Mobil/Exxon SpeedPass® key fobs, some electronic subway fare cards and read-at-a-distance building-access cards, and E-ZPass automated toll-paying transponders.

Power levels vary by application but are typically low (e.g., a few watts in the reader's interrogation signal; milli- or microwatts in a tag's reply. Governmental health and safety regulations for radio signals apply.

RFID tags and their associated readers use various frequencies and encoding schemes. Popular frequencies include 433 MHz, 915 MHz, 2.45 GHz, and 13.65 GHz. In the United States, most RFID systems use unlicensed spectrum at low power under Part 15 of FCC rules.

### *A bit of history*

RFID has its roots in military technology. One ancestor is the "Identification, Friend or Foe" (IFF) system. An IFF transponder, first deployed on military aircraft in World War II, replies to a suitably coded incident radar signal by broadcasting an identification code.

In an early Cold War Russian bugging device called "The Thing," sound waves modulated the reflection of an incident radio signal. The bug emitted a signal (and hence was vulnerable to detection) only while powered by the external source.

US Patent 3,713,148 (1973, Cardullo et. al.) is often considered the immediate ancestor of the modern RFID tag. That patent describes a memory-equipped radio transponder, preferably implemented such that "the transponder generates its own operating power from the transmitted interrogation signal."

Arguably RFID entered the public consciousness in 2004, when Wal-Mart levied an RFID tagging requirement on its top hundred suppliers. Effective January 2005, all cases and pallets sent by those suppliers to the mega-retailer were to have RFID tags.

### *Why not use barcodes?*

Microchip plus antenna cost more than lines of ink. Why not stay with barcodes? For many reasons, it turns out. Consider:

Identifying an object from its barcode involves line-of-sight readout by a laser-based optical scanner. A person individually scans every can, package, and bottle. (Some stores allow shoppers to scan their purchases. That may reflect a sophisticated checkout system, perhaps with scale-based tests to sanity-check the claimed purchases, or trust in the clientele. Regardless, barcode checkout requires that a human handle each item.) In a hospital setting, someone holds a scanner to the patient's barcoded wrist bracelet or test order.

An RFID tag (absent electromagnetic shielding, intentional or otherwise) is polled by an omnidirectional radio interrogation signal. RFID readers that discriminate among concurrent reply messages can identify at once: groceries in a packed cart (checkout application), books in an armload (library application), or pallets in a warehouse (inventory application).

Barcodes are intrinsically surface-mounted labels. An RFID tag can be implanted (an ID tag your cat can't lose), tucked out of sight (barcodes aren't pretty) or otherwise hidden (for example, an unobtrusive theft-avoidance marker).

Lines added to a barcode (e.g., a production batch number appended to a product code) increase specificity; they do not make the barcode smarter. The microchip in an RFID tag can, however, support whole new functions—like a microprocessor. The tag's memory then stores a computer program in addition to the part number.

And memory content need not be frozen at the factory. Many tags use Electrically Erasable Programmable Read-Only Memory, EEPROM. The name is awkward; EEPROM is normally read-only but reprogrammable with suitable equipment. On a production line, individual tags can be rewritten with unit-specific identifiers and manufacturing history. At retail or library checkout, a tag can be rewritten so that it won't trigger the store's or library's theft alarm. (I bet I know what you're thinking. Not to worry, we'll get to privacy concerns.) Updateable tags can capture product experience along the distribution channel; returned items may then give clues to the root causes of breakage problems.

On-chip sensors offer more flexibility. Add temperature sensors, and a cartoon of ice cream can report that it thawed and then refroze in transit. Let a barcode try that trick!

These examples barely scratch the surface of what an RFID tag might do. Memory, microprocessing, sensors, and wireless communication: The possibilities offered by their combinations are numerous, indeed.

### Readout constraints

An RFID tag alone serves no purpose. It must partner with a reader.

The short answer to "From what distance can a tag be read" is: It depends. Many things impact the range and accuracy of readout. These factors include: the power transmitted by the tag (for passive or semi-passive tags, a small fraction of the incident power), the communications protocol between tag and reader, the relative orientation and sizes of tag and reader antennae, interference (perhaps from other RFID tags awakened by the same interrogation pulse), and the sensitivity of the reader itself.

An active tag can be designed to be read from hundreds of feet away. If the active tag uses unlicensed spectrum, its transmission level is constrained by the same regulations as a WiFi local area network or a cordless (not cellular) phone. Passive-tag readouts are more frequently quoted in the range of inches to a few feet. In either case, nothing precludes construction of a more sensitive reader. It takes only the motivation (and money).[1]

PCs get faster, digital cameras hold more images, and TV screens approach life-sized. It's fair to imagine future RFID readers that are more sensitive and less expensive.

And highly portable. Gentag was recently awarded a U.S. patent for an RFID-reading cell phone. [2] An RFID readout can be made and instantly sent—anywhere.

*Detecting* a tag's response and *reading* it are different. Security protocols (signaling methods that authenticate reader and tag to one another) and message encryption can render detectable tag responses unintelligible.

1. The Jupiter-orbiting Galileo probe transmitted about twenty watts. The signal intercepted by one of the seventy-meter dish antennae of NASA's Deep Space Network had attenuated to about ten zeptowatts. (One zeptowatt = one sextillionth ($10^{-21}$) of a watt.) That's not to say anyone is apt to use DSN for scanning RFID tags.

2. http://www.gentag.com/IP%20Release%2020050088299%20A1.pdf

### *Processing system*

Things get interesting when we consider vast numbers of related RFID tags, whether elevator-control cards for the workers of a secure office building, E-ZPass transponders for East Coast commuters, or crates in a Wal-Mart warehouse. When related items bear RFID tags, there is, behind the scene, a data repository that enumerates those items. Another repository saves the tags' readouts. A computer manages those repositories.

(Side note to those with programming backgrounds: RFID tags used for a common purpose employ a common format. For many purposes, the collected data from similar tags are most efficiently compiled and searched within a simple file system. More complex applications may ingest tag-related records, or copies of them, into a general-purpose—but more compute-intensive—database management system. A DBMS adds value when the goal is to exploit relationships between RFID data and other relevant information like customer accounts and employee files.)

Consider managing access to a building through machine-readable employee badges. The system tests whether a sensed ID belongs. (Data look-up: Is this a valid access card?) It determines whether the badge requires special handling. (Data look-ups: Has this badge been reported lost? Are access times restricted for this employee?) It annotates when the employee entered the building (data capture).[3]

Thousands of workers report daily to that secured office. Thousands of cars with E-ZPass transponders pass toll plazas daily. Thousands of Nike sneakers roll daily off the assembly line, each bearing a unique RFID. Badges, transponders, and sneakers: Somewhere, there is a data repository about each.

Tags, readers, and computer/repository are the components of any RFID system.

---

3. This description applies whether the badges use RFID or magnetic-stripe badges and readers.

### *Avid for RFID?*

Does RFID sound great? Then it's time to revisit our opening scenarios.

Recall three basic RFID properties. Tags are read from a distance. Tags may be personalized. Tag transactions flow to a data repository, where they can be combined, contrasted, and correlated with other information. Together, these features explain our incidents.

First, the peanut alert. The bakery assigned a unique number to each batch of cookies. The grocery, on receiving the batch-specific recall notice, searched its sales history. You once joined the store's loyalty program to qualify for discounts; the store matched your shopper ID (remember the key fob they gave you?) with a cookie label (identified by its RFID tag). That's how they can contact you. (This scenario might also occur with barcode-identified cookies, but that's less likely. RFIDs better accommodate customization, especially on small production runs, than do printed labels.)

Next, the tire recall. The manufacturer lost track of many customers over the years. Whether altruistically or for fear of a lawsuit should a tire blow out, the manufacturer contacts toll-road authorities across the country. Each tire has an embedded individualized RFID tag. You commute on a toll road; your car's tires, like your toll-paying transponder, have been noted. The toll-road authority has a database of transponder account holders; it sends (sells?) your address to the manufacturer.

(A case can be made that's a good thing. Individually labeled tires make sense. A tire may fail for reasons related to the car model on which it was mounted, the factory in which the tire or car was built, or improper handling anywhere along the distribution channel. Tires from the Firestone Decatur plant figured most prominently in the Ford Explorer tire recalls of 2000. Compared to the cost of a recall, a few cents per tire for a reprogrammable tag with which to capture tire-specific detail could prove a sound investment.)

Next, the computer heist. Your employer's access-management system relies on RFID-enabled badges. A crook loitering in the lobby with a wireless "sniffer" in his briefcase captured the signaling between your badge and the door-lock control. That's sufficient information to get into the building, looking to the security system like you.

(To be complete: For any reasonable access-management system, the spoofing trick demands greater sophistication than mere record and playback. The tag/reader wireless link will be protected by security protocols and encryption. Still, systems do get cracked. As I write, in August 2006, several RFID-based e-passport systems have been reported hacked.)

Finally, your abduction wasn't random. Your new passport, which contains an embedded RFID tag for faster authentication and to deter counterfeiting, enabled terrorists with portable RFID readers to identify you as one of the foreigners in the crowd.

### *Could you be any more paranoid (or unrealistic)?*

The peanut warning was to your benefit. You might have had a peanut allergy. It's hard to feel warm and cuddly about the other scenarios.

But wait, you say. These aren't *realistic* scenarios. Data are encrypted, to prevent eavesdropping. Databases are secured. Laws bar disclosure of personal data. The sheer quantity of responses from countless myriads of tagged garments, vehicles, pets, whatever, collected at thousands of readers, spread across the country and the world, accumulating day after day after day … surely those oceans of data render *my* digital presence effectively invisible.

I hope you're right—but I see too many counterexamples for optimism.

### *Encryption*

If the faint signal from most RFID tags doesn't prevent eavesdropping (but can't miscreants build more sensitive readers?), encryption might protect us. The argument goes: If eavesdroppers don't understand what a tag says, we're okay.

What's the track record of encryption-based security in high-volume consumer products? Do security-enabled products remain secure?

No.

Digital video disk (DVD), satellite TV, and cable TV use encryption-based protection schemes. All have been cracked. In the United States, the Digital Millennium Copyright Act, 1998, outlaws such activity and discourages many—but certainly not everyone—from viewing content without paying, or from copying content without authorization.

RFID tags have cost and power constraints (read: computational limits) far more stringent than DVD players, satellite receivers, and TV set-up boxes. How credible is it that the security mechanisms in RFID systems will resist cracking?

Encryption schemes transform unencoded data into a new representation. The value of encryption lies in the difficulty (in the absence of presumed secret knowledge—the decoding key) of reversing the transformation. Encryption schemes fail when the secret is disclosed, the algorithm proves less mathematically challenging than believed, or technological advances make once impractical transformations easily computable.[4]

Recall the Data Encryption Standard (DES), a US federal standard introduced in 1976. DES uses a 56-bit key. DES was, at the time of its introduction, trusted to secure banking transactions.

Now fast-forward to 1998, when the Electronic Freedom Foundation broke DES in less than 72 hours with a custom DES cracker built for under $250,000. In 1999, a band of computer enthusiasts cracked DES with an ad hoc collection of personal computers in under twenty-four hours.[5] Today, we use *128-bit* encryption to buy books from Amazon.[6]

---

4. Secure systems rely on encryption not only to obscure messages, but also to authenticate sender and receiver to each other. After all, a "secret ID" broadcast in the clear isn't much of a secret.

5. http://www.networkworld.com/news/1999/0120cracked.html

6. 128-bit encryption is the US domestic standard for web browsers. Export versions of browsers have weaker encryption, but are readily upgraded.

What accounts for code breakers' progress? More than anything else, Moore's Law.

Dependably since Gordon Moore, co-founder of Intel Corporation, articulated this forecast in 1965, the number of transistors on an integrated circuit chip has roughly doubled every two years. Data densities more or less double in eighteen months. Simply put, Moore's Law suggests that in five years computers will be five or six times more powerful than today, with ten times the storage. Hackers *will* have more powerful PCs, but many five-year-old RFID tags will remain. (My shoes don't last five years; sometimes my tires do.) For how many RFID-tagged possessions will encryption be cracked while I own the item?

### *Sometimes encryption Isn't enough*

Several central banks are considering embedding RFIDs in new currency. Embedded RFIDs will make counterfeiting harder. With a simple *ping*, a sales terminal can tally the cash in a drawer, and a bank can confirm the money in its vault.

And with a simple *ping*, a mugger can find the people in a crowd carrying wads of cash. The mugger does not need to know the serial number on each bill.

A unique, machine-readable ID on a bill raises other issues. Imagine an ATM record of a specific $20 bill, tied (of course) to the account of the customer who withdrew it. Next, imagine that bill appearing in a cash drawer, concurrent with a specific purchase. At least in theory, RFID-enabled currency makes cash purchases traceable.

On the bright side, if money becomes associated with its owner, cash reported as stolen will announce itself to the police (not in real time, but soon after it's spent, when the business makes a cash drop at its bank). Thieves will be wise to spend their loot quickly (before the theft is reported and the serial numbers of stolen bills reconstructed from widely distributed records) and/or at businesses without RFID-reading registers.

The RFID reader in a cash register must discriminate currency tags from other tags (say, the tags embedded in high-value coupons also in the cash drawer). The distinguishing data patterns that make cash recognizable as cash *will* become known. Assume no one who knows the secret algorithm talks, whether for personal gain or under duress. Assume the algorithm cannot be cracked. The secret to recognizing currency by pinging wallets will become known soon after the first cash-counting register is stolen.

When RFID-tagged money enters circulation, expect foil-lined wallets to follow.

(December 2011 update: Common sense prevailed. To my knowledge, no country has introduced RFID-tagged currency—but the concept lives on. Beware RFID-tagged casino currency.[7] Exchange your winnings as soon as possible, or keep the chips in a foil-lined sack. Someone with a portable RFID reader could be scanning for high rollers.)

### *Spoofing*

The ID in RFID stands for identification. Mimicking or copying a tag means forging credentials, with all the opportunity for mischief a false identity implies.

For simple systems, simply recording and playing back the automatic response of an RFID tag will fool a reader. For applications using robust encryption, a spoofing device can be constructed after the security scheme is understood. And there's always cloning. A cloned RFID tag *will* fool an RFID reader.

(Here's an exception: Cloning alone may not suffice if the RFID tag provides a second identification mechanism. We've seen that the microchip in an RFID tag can implement other functions, including sensors. A biometric sensor initialized to recognize my fingerprint should stymie someone not me using a clone of my ID tag. The RFID spares me from sharing a fingerprint scanner—oh, the horror—by making the access system much costlier.)

Does cloning of RFID tags work? Yes. It's been demonstrated at the Black Hat hackers conference (everyone has conferences!) with

7. http://en.wikipedia.org/wiki/Casino_token

prototype RFID-enabled US passports.[8] (In a bit of good news, the U.S. State Department has responded to security concerns. As finalized, the new RFID-carrying U.S. passport includes metal sheets in its covers and extra security software to control access to the onboard personal data.[9] Of course it's only a matter of time until someone steals and clones the border officers' special passport reader.)

RFID spoofing offers new ways to gain unauthorized access to a building (or country), create an alibi, and plant evidence implicating another.

### *Misused data*

Might heretofore invisible traces of your activities now be used to your detriment? Our answers may differ.

(It's not that the opportunity for data misuse is unique to RFID. To the extent RFID tags record your life in greater detail than was once practical, data misuse becomes that much more problematical.)

Appropriate bounds on government data collection is a hot topic as this is being written: Warrantless wiretaps. Monitoring of international money transfers. Sifting of domestic call records (who calls whom, not phone taps). Prescreening passengers against a No Fly list.

A strong antiterrorism case is made for such searches. A strong civil-liberties case is made against them. It's worth noting that these controversial, data-intensive searches involve government accessing businesses' databases. If you have an opinion about government using data gathered by your phone company, bank, and airline, you probably also care about the repurposing of RFID-centric repositories.

Will companies and public-transit agencies agree—or be required—to share their RFID records? If asked, will institutions (say, gas stations) tweak their RFID reader software to accept responses (say, from tires) previously discarded as irrelevant? Will law enforce-

---

8. http://www.technewsworld.com/story/52270.html

9. "Passport Chip Sets Security Concerns," *The Wall Street Journal*, August 9, 2006.

ment and antiterrorism agencies—or bad guys—deploy their own RFID readers across the land? If those questions concern you, the proliferation of RFID tags should also concern you.

It is certainly plausible that government will take interest in new ways to track the activities and movements of individuals.

### *Lost data*

All this data sniffing, code cracking, and spoofing sounds so ... James Bond. It can't possibly apply to John Q. Public. Surely those digital footprints will be secure within corporate and governmental databases.

History calls into question the ability of organizations to protect their data.

ChoicePoint is a low-profile corporation that aggregates and sells personal data about most of the U.S. population. Its 19 billion records (as of February 2005) include social security numbers, credit reports, and bank-account and credit-card information. In late 2004, at least 100,000 people's records were exposed to hackers.[10]

Consider these 2006 news items. AOL accidentally released data about nineteen million searches by 650,000 users. The searches reveal social security numbers, among other items most people tend not to share. The Department of Veterans Affairs lost a laptop whose hard drive stored personal and financial data for more than twenty-five million veterans, active-duty military personnel and spouses. Ernst & Young, a "Big Four" accounting firm, lost data on more than two hundred thousand Hotels.Com customers through a laptop theft.

(And in 2011? Computer security seems only to have gotten worse.[11])

To me, trust in the infallibility of data-center security seems misplaced.

---

10. http://www.consumeraffairs.com/news04/2005/choicepoint.html

11. I write regularly about privacy and network (in)security topics on my authorial blog, SF and Nonsense. A June 2011 post that reviews several recent incidents is "Hackpocalypse Now" (http://edward-m-lerner.blogspot.com/2011/06/hackpocalypse-now.html).

***Data overload?***

Perhaps the sheer amount of data will obscure our RFID tracks.

The numbers involved seem daunting. Stores, warehouses, and toll plazas will each have multiple RFID readers. In large cities, it's easy to imagine hundreds or thousands of readers per square mile. Readers can take two hundred readings per second. (A reader which polls that frequently may detect the same tags [or no tags] over and over. Filtering software will discard repetitive data that add no value.)

How much data are we talking about?

While the least expensive tags will report only a few bytes, a payload of 2-3 kilobytes is more typical.[12] To the tag's own data payload, the reader and/or the associated data system will then append metadata (such as a timestamp and the reader's own identifier or location).

It's easy to construct a comforting scenario. Big Brother databases require continuous coverage across large areas over long time periods. Is it practical to retain that much data?

As a thought experiment, consider a future retail district. Within one square mile we find fifty large RFID-enabled businesses, each with one hundred RFID readers. (If you prefer, make that five hundred small businesses, each with ten readers.) That's five thousand readers per square mile. Retain one record per reader every ten seconds throughout an eight-hour business day.

(Reminders: Many RFID applications are intentionally short-range—at checkout we want data from the shirt the customer selected, not the thousand shirts he didn't. Tags ignore pulses of the wrong frequency, too low a power level, or using the wrong protocol. Upstream data systems discard tags deemed irrelevant, such as sequences of unchanged readings.)

That's 14,400,000 (= 5,000 * [0.1] * [8 * 60 * 60]) readings. At 3 kilobytes of data per reading we get ~43 X $10^9$ (that is, 43 gigabytes) of data in eight hours, from that square mile.

A top-end iPod holds 60 gigabytes. Apple sells them by the millions.

---

12. http://www.rfidjournal.com/faq/18/66

What if 50 GB were collected daily in every populous square mile across the United States? Most of us live in cities, which also seem to be the favored terrorist targets.

The nation's area is about 3.5 million square miles. One tenth that area encompasses 140 disjoint regions each of fifty miles square (2,500 square miles). That's many cities, and most U.S. residents.[13,14]

Our thought experiment has expanded to (50 gigabyte/$mi^2$) * (350,000 $mi^2$).

This *is* a large number: 1.75 X $10^{16}$ bytes (about eighteen petabytes) daily.

Organizations do deal in petabytes. Equifax, the credit-reporting company, faced such numbers in 2002.[15] Then there's the Large Hadron Collider, nearing completion at the European nuclear research center (called CERN from its French acronym). The LHC computing grid, a globally dispersed, multi-supercomputer, will

---

13. Deploying 5,000 readers per square mile across such a large area is futuristic. It entails 1.75 billion readers. Estimates vary, but today's worldwide RFID *tag* population is probably about one billion.

    December 2011 update: The future keeps rushing at us. By 2009, worldwide *annual* RFID tag sales exceeded two billion units (http://www.prnewswire.com/news-releases/reportlinker-adds-rfid-forecasts-players-and-opportunities-2011-2021-101460354.html). RFID tags in apparel alone may reach 20 billion units annually by 2021 (http://linux.sys-con.com/node/2009877). Although I've not found population estimates of RFID readers, I see no reason to doubt that the reader segment of the industry has also experienced rapid growth.

14. http://www.census.gov/geo/www/mapGallery/images/density90.jpg illustrates (with 1990 census data) a very concentrated U.S. population. Ten percent of the area certainly covers most of the population.

    2011 update: The corresponding graphic based upon 2010 census data doesn't look much different: http://www.census.gov/geo/www/maps/2010_census_nighttime_map/nighttime_map_2010.html.

15. The source data was http://www.cio.com/archive/101502/bigger.html , but the cited page has since vanished. Here's a snapshot from the Wayback Machine: http://web.archive.org/web/20080906192535/http://www.cio.com/archive/101502/bigger.html.

store 15 petabytes per year.[16] As a third example, the San Diego Supercomputing Center recently acquired an 18-petabyte tape library to support its work in Earth sciences.

Data centers large enough to accept daily torrents of RFID data are not farfetched.

(December 2011 update: such data centers are *here*. LHC and its repositories are online. Stanford Linear Accelerator Center is planning a 100-petabyte archive of astronomical imagery from the Large Synoptic Survey Telescope.[17] And we learned in August 2011 that IBM Research had built a 120-petabyte storage system for "an unnamed research group."[18]

Looking ahead a few years to the construction of the Square Kilometer Array, a new radio telescope, IBM developed a software architecture to handle the SKA's expected *daily* data output of more than an exabyte—that's a thousand petabytes.[19])

16. "Old World, New Grid," *IEEE Spectrum*, July 2006 issue. Somewhere in all that data, it is hoped, will finally be found evidence of the Higgs boson (a missing piece from the Standard Model of particle physics).

Update: On December 13, 2011, researchers announced that the elusive Higgs boson *may* have been sighted. Alas, the few tantalizing glimpses—among trillions of proton-collision events observed at the Large Hadron Collider—remain far from the level of statistical significance expected of a scientific finding. And so, the *Washington Post* reports, "Rolf-Dieter Heuer, director general of the European Organization for Nuclear Research (CERN), home of the Large Hadron Collider (LHC), said another year's worth of data needs to be compiled before anyone can reach 'a definitive answer on the Shakespearean question on the Higgs: To be or not to be?'"

See: http://www.washingtonpost.com/national/health-science/scientists-say-they-are-homing-in-on-higgs-boson/2011/12/13/gIQAsqcnrO_story.html

17. https://news.slac.stanford.edu/features/slac-led-project-build-worlds-largest-digital-camera-impresses-doe-panel

18. http://techcrunch.com/2011/08/26/ibm-assembles-record-120-petabyte-storage-array/

19. http://www.scoop.co.nz/stories/SC1111/S00063/ibm-completes-software-project-for-square-kilometre-array.htm

### *Processing overload?*

Okay, it's possible to store all the data, if not today, then surely after a few more Moore's Law doublings. Who can possibly locate anything within it?

Consider an iPod. Finding a tune does not require listening at random to thousands of songs. You search an organized index, whether by title, album, group, or performer. The indices were created and are continually updated as music files are uploaded from your Mac or PC.

So, too, will RFID tag records be labeled and organized for expedited search as they are uploaded from RFID readers.

Our thought experiment assumed archiving of three kilobytes per RFID reading, but it's not all needed for searching. Many queries will involve only time-and-place values or an ID code (of specific tags or associated persons). Everything else will be retrieved after the record(s) of interest has been found.

Time and place sorting are effectively free. Businesses and local governments (like toll-road authorities) will own most RFID readers. They will preprocess their tag readouts for their own purposes. The data are intrinsically geographically presorted, simply by knowing which reader(s) read what tag. The records are also intrinsically time-sequenced.

RFID data can thus be searched efficiently within and among organizations. Knowing the locale or timeframe to search quickly pares the amount of data to be examined.

We're contemplating a great deal of data, but some organizations are in exactly that business. Take two examples. Equifax, in the 2002 article cited above, describes sorting six months of data, 67 billion database records, in three seconds. On August 29, 2006, the Federal Bureau of Investigation demonstrated its Investigative Data Warehouse.[20] IDW reportedly contained on that day 659 million records among fifty databases. The FBI spokesman said a search of a thousand names and birthdates that in 2002 would have required

20. "FBI Shows Off Counterterrorism Database," *The Washington Post*, August 30, 2006.

more than 32,000 hours can now be done within thirty minutes.

Moore's Law again.

### *Parallel processing*

But not only Moore's Law.

Large-scale computing challenges often yield to a divide-and-conquer approach—what computer scientists call parallel processing. It's true of climate studies (many small atmospheric volumes modeled concurrently), H-bomb simulations (far tinier volumes, in much shorter time slices), and data mining (subsets of vast data aggregations). Specialized computers harness many processors in parallel to more speedily calculate, sort, and search. Such computers gain increased throughput with faster chips or increased parallelism.

Searching RFID repositories is a naturally parallel application.

IBM just received a U.S. Department of Energy order, for 2008 delivery, of a highly parallel supercomputer.[21] The aptly named Roadrunner will contain more than 16,000 commercial processors: AMD's Opteron chips, common in servers, and IBM's Cell chips, soon to be found in Sony's PlayStation 3. Roadrunner will achieve a throughput of $10^{15}$ operations per second.

Roadrunner appears well suited to exploiting even the largest RFID repositories.

### *The miracle we forget*

Google's cofounders began research on a search engine in 1996. By 2005, the *New York Times* reports, Google had indexed eight billion web objects, with more being added continually; the aggregated data volume in those objects is vastly larger. Constantly maintained indices enable us to search across those billions of items, often within a fraction of a second.

Web objects exhibit far more variability (read: they're harder to parse and index) than RFID records. Web objects include text pages, images, Usenet messages, and linked files of various types. It all keeps growing, of course. Google aspires to scan and make

21. "U.S. Taps IBM for Supercomputer," *The Wall Street Journal*, September 7, 2006.

searchable *every book ever published* (although copyright issues may constrain that ambition).[22]

And Google does most of its work with swarms of commodity PCs.

In summary, precedents exist for handling data collections arguably of the size of a useful national RFID data repository. Moore's Law assures us that the ability to collect, sort, associate, and search data will increase. The proof is as near as your favorite search engine.

### *Did we have privacy anyway?*

Some would argue that RFIDs are but one threat to our privacy.

You can be located by the mobile phone in your pocket (which is a good thing in association with a 911 call). You can be tracked by a GPS transmitter hidden on your car or in your child's teddy bear (GPS-enabled stalking has been reported). You may work somewhere, like downtown London, strewn with closed-circuit TV cameras (which were invaluable in investigating the 2005 London subway bombings.). Millions of people carry camera phones.

Perhaps privacy *is* a thing of the past. Privacy was a recent phenomenon anyway. It doesn't exist in small communities—where most people lived, for most of human history. Privacy didn't exist on telephone party lines.

Some (the author among them) still regret the level of surveillance to be made possible by the confluence of RFID data collection and data mining.

### *And yet ...*

Public law may yet save us from Moore's Law.

There are a few encouraging signs that the privacy risks of RFID tags, especially coupled with data mining, are entering civil discourse. Public interest groups, such as the Electronic Frontier Foundation and the American Civil Liberties Union, are engaged.

22. http://www.nytimes.com/2005/09/21/technology/21book.htm

Industry groups increasingly acknowledge public concerns. Some in government recognize the potential problem, as noted in recent GAO testimony to Congress.[23]

In June 2006, U. S. Senators Dorgan and Cornyn proposed to their colleagues the formation of an RFID Caucus; its scope would include (albeit as the final item in their list) "privacy and security."

The privacy-motivated redesign of the RFID-equipped U.S. passport was certainly a positive step.

States are filling the policy void left by federal inaction on RFIDs and privacy. (States took a similar leadership role a few years ago, in the absence of federal data privacy standards. Many states modeled their legislation on California's privacy law. Much that we've recently learned about lost and stolen personal data comes from state-mandated reporting.) On August 31, 2006, the California legislature passed the Identity Information Protection Act, setting privacy standards for RFID tags in state identification cards.

If the governor had signed the legislation, it would have been among the first RFID-centric legal protections. Instead, he vetoed the bill, deferring to future federal standards.

(2011 update: RFIDs and RFID-originating databases remain largely unregulated.)

How might an informed citizenry choose to influence RFID legislation? To name a few aspects: Disabling or removal of RFID tags in consumer goods at the point of sale. Transparency as to the intended uses of collected data. Rules to limit tracking people through their RFID traces. Standards for the secure transmission and storage of data reported by tags.

Do you have opinions about RFIDs? If so, consider sharing them with public officials. http://www.firstgov.gov/ offers contact information for office holders at all levels of government.

---

23. http://www.gcn.com/online/vol1_no1/40817-1.html

### SF implications

Dark futures and dystopias abound in science fiction, and RFIDs offer a new flavor of doom. I explore a few possibilities in my story "The Day of the RFIDs."[24] The hero foresees many of the dangers outlined above, and strives to prevent a Big Brother outcome.

Please join me in wishing him luck.

How might RFID technology influence our fiction?

It's easy to foresee a time when little can be accomplished in secrecy, and less can escape after-the-fact discovery through data mining. Still … human nature (and story plots) demand that some of us will rebel. How, in an RFID micro-monitored future, will people conspire, commit crimes, or, in a pinch, assume a new identity? Neither forged paper IDs nor counterfeit RFID tags will suffice; any credible alias must involve extensively altering the archives of past RFID readouts.

RFIDs implanted under the skin could mean never lost, never left at home pointers to our medical history. In an emergency, that information could literally be a lifesaver. Easy access to medical-record identifiers also risks a staggering loss of personal data, and with it, individualized discrimination. If any healthcare professional can access your files, how secure can that data be?

Many RFID tags are programmable. So are many RFID readers, facilitating product upgrades. Programmability opens the door to hacking, computer viruses, and other malware attacks at every point from tag to archive. What are the opportunities for vandalism, theft, forgery, impersonation, and cyber-terrorism?

RFID readers will be spoofed, stolen, and cloned. How much RFID data will be misappropriated or manipulated, and to what ends?

Imagine a society microscopically observed via RFIDs. Now something, perhaps a time-delayed virus insinuated into a region's

---

24. In the collection *Creative Destruction.*

tag readers, corrupts the data stream and eliminates accountability. How might a city with a big hole in its data history reintegrate into the larger economy?[25]

RFIDs *are* coming. We can at least hope for interesting fictional escape.

### *Some websites of interest*

#### *From industry sources:*

http://www.rfidjournal.com/article/articleview/1336/1/129/

http://www.rfidjournal.com/faq

http://www.technovelgy.com/ct/Technology-Article.asp?ArtNum=50

#### *About privacy issues:*

http://www.cdt.org/privacy/20060501rfid-best-practices.php

http://www.rsasecurity.com/rsalabs/node.asp?id=2120

http://web.archive.org/web/20110606021953/http://w2.eff.org/Privacy/Surveillance/RFID/

25. Soon after this essay saw print, I tackled that question in a story. "The Night of the RFIDs," technically a sequel to "The Day of the RFIDs," but quite standalone, is part of the collection *Countdown to Armageddon / A Stranger in Paradise.*

# Where Credit Is Due

*"Give me a one-handed economist! All my economists say, 'On the one hand … on the other.'"*

— Harry S. Truman

Winslow Justinian Prescott waited outside the packed pressroom. The cacophony within was extraordinary, the White House correspondents variously suspicious, curious, and baffled. They knew something momentous was afoot. He had not held a press conference for months.

Prescott adjusted his tie, flicked a bit of lint from the lapel of his suit, and stepped inside.

"The President of the United States," a booming voice intoned.

The noise stopped like a switch had been thrown. Few of these correspondents liked him. Some didn't even respect him—but by God, they all respected the office. He had the best job in the world, and he meant to keep it. After his announcement—all the more dramatic as an October timeout from the campaign trail—reelection was in the bag.

Prescott stepped behind the podium that bore the presidential seal. "I have a statement, and then I'll take questions."

Text appeared on the teleprompter. "Ours is a vibrant and sophisticated economy, powered by 350 million complicated people. And America is a cog in the larger world economy, driven by eight billion people." *Subtext: The recession is no one's fault.* "It's not surprising the economy sometimes performs less well than we would hope. The wonder is that anyone dares try to guide it."

Before him, styli scribbled, cameras stared, and solid-state recorders did—well, nothing discernable. He kept reading. "1776 will ever be associated, and rightly so, with America's declaration of independence, but a second revolution also began that year." In the holo display beside the podium, a stolid and bewigged figure appeared. "In March of 1776, Adam Smith, the father of modern economics, published this book."

A leather-bound volume, its title emblazoned in gold leaf, replaced Smith. *An Inquiry into the Nature and Causes of the Wealth of Nations.*

Reporters squirmed.

*He'd* told *the speechwriters to cut the history lesson. He was the president, not class know-it-all.* "Many economists followed Smith. They found things to agree upon—and more not to."

Grins throughout the crowd. Politicians, lawyers, and economists … no one objected to a potshot taken at any of them.

"They weren't *entirely* wrong." A few chuckles. "Every one was right—about some of us, some of the time. What have economists asserted? People make rational decisions." His raised eyebrow drew another laugh. "People value money above all else." If that were true anywhere, DC wasn't the place. Here the pursuit of wealth ranked a distant second to the quest for power. "Taxes are too high *and* too low."

The squirming had stopped. He had them in the palm of his hand.

"No psychologist would claim people are always rational. She'd be out of business." Laughter. He skipped a riff about sociologists. Always leave your audience wanting more.

"It's no one's fault. It's hard to know even our own minds. We wonder what our significant other or child or neighbor might do minutes from now, yet we expect economists"—and politicians—"to predict how we'll all act and react for years to come."

*Harry Truman, prepare to stand corrected. The buck* won't *stop here.*

In Prescott's mind's ear: a theatrical drum roll. "People are rational and rationalizing, methodical and impulsive, principled and expedient, wise and whimsical, self-interested and selfless. People are … human."

Fast-forward through economists claiming mercantilism, whatever that was, and capitalism, and socialism, and communism were the way to go. Past supply side and demand side, past voodoo economics and the Gross National Happiness. "Our economy became too complex for any of us to comprehend. But *this* administration has the solution."

*That* made everyone sit straighter.

"Greedy or generous, rich or poor, playing now or saving for later … who can calculate all the possibilities? Economics is a job for someone who *can* understand it all. All our needs, all our moods, all our aspirations—all together, all the time. Economics is a discipline grown too large for the human mind to grasp. It's become a task for an *artificial* intelligence."

"My experts assure me …" Like all experts, they *assured* him of nothing. Every remark sagged beneath the weight of caveats. They whined about more supercomputers. Extra data storage. Additional statistics as inputs. Above all, they wanted yet more testing.

If technology were the solution, then past techie dawdling was the problem, and he needed to take credit now, not after the election.

Well, the presidency had its perks. Few dare say no to you; fewer could make it stick. Certainly very few civil-service scientists and programmers.

Prescott skipped to the end of his prepared statement. The ancient tome shimmered and was gone. A virtual desktop took its place. "It's time to introduce the father of a *new* era in economics." *Delivered by the leader to whose stewardship your readers and listeners and viewers will entrust the next four years.* "I give you … the economist who thinks like us all."

He rolled the podium trackball over an unlabeled icon and clicked.

Laughter erupted—but this time they weren't laughing *with* him.

Something moved in his peripheral vision. Prescott turned toward the scrolling words.

*You have reached the computing array of AIdam Smith. I am taking a sick day. Please leave a message.*

The shouted questions began ….

# *Of Space, Time, and Quantum Weirdness*

Are you ready for a change of scenery?

The universe is *so* huge. Wouldn't it be nice if we could experience it all? Whether or not we have the place to ourselves, what wonders we might behold! Black holes and supernovae. Binary stars. Worlds of every imaginable size and climate (and, perhaps, biological novelty).

Science, alas, tells us that the universe has a speed limit. That speed limit is *way* too slow for mere humans to explore our galaxy, let alone to reach other galaxies. A century of experiments and observations have borne out Albert Einstein's theory of Special Relativity.

Until in September 2011, one credible experiment didn't.

The news that neutrinos might, just possibly, just barely, travel faster than light broke—naturally—mere days after the magazine issue with "Faster Than a Speeding Photon" went to press. And so, the essay in this collection that I would have *least* expected to need updating, did.

Will faster-than-light (FTL) neutrinos be confirmed by independent research? Fond wishes aside, it's worth remembering that for several centuries, experiments and observations supported Isaac Newton's theories, too. Newton wasn't mistaken, only incomplete.

In short: see "Faster Than …" for a look at FTL concepts, in science fiction and from the frontiers of science, to see how an FTL breakthrough might happen.

Because any discussion of relativity deals with time as well as space (more precisely, with four-dimensional space-time), the essay also touches upon time travel.

Einstein has more than relativity to his credit. His Nobel Prize for

Physics recognized unrelated work that helped lay the groundwork for understanding the universe at atomic (and smaller) levels. In a phrase, for quantum mechanics.

Quantum mechanics is, alas, highly weird. How weird? Consider an electron that (a) passes through a partition with two slits and (b) strikes a screen behind the partition. Is the electron a particle that passes through one of the slits? Or is the electron a wave that passes through both slits, in the process creating an interference pattern on the screen? The trick answer of QM—borne out by experiments—is: it depends on how you observe the event.

(If you're unfamiliar with quantum mechanics or find the topic incomprehensible, consider this from physics Nobelist Richard Feynman: "I think it's safe to say that no one understands quantum mechanics." That said, the math of QM correctly describes the subatomic domain. In the essay I'll touch upon some of that.)

Einstein's Theory of General Relativity provides a (so far) flawless description of the large-scale structures and behaviors of the physical universe, governed by gravity. Quantum mechanics describes the quite different behavior of the exceedingly small, at a scale in which gravity seldom matters. A foundational conflict within modern physics is that General Relativity and Quantum Mechanics make wildly incompatible assumptions about the nature of reality.

The universe, of course, is indifferent to our tidy notions of compartmentalization. There's no boundary at which the universe switches from obeying one theory to obeying another. Very awkward, that: we need GR to describe the gravitational effects of, say, a black hole—and yet a black hole is a quantum-sized object.

(That is: given enough mass, gravity collapses it all into a dimensionless point. When you see mention of a black hole's size, that's a reference to the "event horizon" that surrounds the black hole. The more mass in a black hole, the larger its event horizon.

Should you have the misfortune of falling toward a black hole, the event horizon is the point of no return. Fall inside the event horizon and you'd have to break the universe's speed limit, the speed of light, to get back out.

Of course, you're on the verge of reading about if/how one can do just that.)

To understand something as massive as black holes (which form, astronomers tell us, in the death throes of stars at least five times as massive as our Sun), we must also resort to quantum weirdness.

Here's the good news: whatever the revolution in physics that supersedes and subsumes GR and QM is apt to give us wondrous new technologies. Such as, perhaps, faster-than-light travel.

In one way or another, each of the four stories that conclude this section relies upon physics at the cutting edge.[1]

1. "At the Watering Hole" relies on FTL travel, and I considered including that short story in this section. But the story is *about* interspecies communications; that's why you encountered it earlier, in the section "Are We Alone? (Life, the Universe, and Everyone)".

# *Faster Than a Speeding Photon*

## *The Why, Where, and (Perhaps the) How of Faster-Than-Light Technology*

(First published in January 2012)

In an earlier essay, I discussed literary tropes and their science-fictional variant. As a memory jogger, a literary trope is a theme, such as heroes and quests, for which the storyteller relies upon preexisting audience familiarity and acceptance. Many an SF trope uses—and advocates of the mundane SF movement would say, abuses—some aspect of science in a nonliteral way.

This essay looks at travel and communications at rates faster than the speed of light. Einstein's theory of special relativity (SR), a piece of very settled and successful science, states can nothing can travel faster than light. More accurately, SR states that nothing can go faster than light in a vacuum; hereafter that qualification is implied. (Even that isn't exactly what SR says, but we'll begin there.)

There's a lot of FTL technology in science fiction.

Among novel series with interstellar scale and reliant upon FTL travel are the Skolian Empire (Catherine Asaro), Uplift (David Brin), and the Lost Fleet (Jack Campbell), and that's off the top of my head, from the very beginning of the alphabet. On the TV and movie front, major franchises dependent on FTL travel include Star Trek, Star Wars, and Stargate.

Often that FTL technology is a trope. But not, as we shall see, always.

### *A matter of distance*

Before reviewing the science, conventional or speculative, let's see why FTL appeals to the storyteller. In a word, because the universe is *big*.

Suppose I wish to set a story on a planet of a star other than our sun. The setting nearest Earth is the three-star Alpha Centauri system. Alpha Centauri C (aka, Proxima Centauri) is a red dwarf star ~4.22 light-years (LY) from Earth; its companions are a bit farther at 4.36 LY.

That 4.22 doesn't sound far, but that's because our unit of measure, the LY, is so vast. Traveling at about 300,000 km/sec (186,000 miles/sec) in a vacuum, light goes about 10 trillion km (6 trillion miles) in one year. Using more precise numbers (9.46 trillion km/year) the distance to Proxima Centauri is ~39.9 trillion km. In scientific notation, that's $3.99 \times 10^{13}$ km.

A *long* way to move my characters.

For comparison, consider "merely" the distances that separate Earth from more familiar objects.[1] The average distance between the Earth and its moon—the limit, to date, of in-person human exploration—is only $3.84 \times 10^{5}$ km. The average distance from the Earth to the sun, also known as one astronomical unit (AU), is ~$1.5 \times 10^{8}$ km. From the sun to Mars, the average distance is ~$2.28 \times 10^{8}$ km, or 1.52 AU. (Earth and Mars approach each other more closely, on the order of $5.5 \times 10^{7}$ km [or 0.37 AU.]) From the sun to Pluto—yet to be visited by any human-built space probe—is, on average, $5.9 \times 10^{9}$ km (39.5 AU).

Thirty-three years after launch, NASA's *Voyager 1* probe—the fastest-moving object ever launched from Earth, having gone farther

1. For distances and much more information about these and other astronomical objects, see http://nineplanets.org/ (about our solar system) and http://www.solstation.com/stars.htm (about nearby stars).

than anything else ever built by man—is ~17.4 billion km (116.5 AU) from the sun.[2, 3]

Let's contrast *Voyager 1*'s epic exploration with the distance separating the sun from next-door neighbor Proxima Centauri. That's (17.4 X $10^9$ km / 3.99 X $10^{13}$ km) ~ 4.36 X $10^{-4}$. *Voyager* has traveled only a small fraction of one percent of the distance to the nearest star.

Past nearby stars, the numbers get yet more daunting. The Milky Way Galaxy—and who among us hasn't a soft spot in his heart for sagas of galactic empires?—is, loosely speaking, a disk measuring 100,000 LY in diameter and 1,000 LY thick.[4]

### *Storytelling with and without FTL*

There are great science-fictional stories that embrace these challenges in distance and time. Some involve generation ships, like Brian Aldiss's *Starship* (in the UK, *Non-Stop)*, Harry Harrison's *Captive Universe*, and Robert A. Heinlein's novella "Universe" (which became the opening segment of *Orphans of the Sky*). Other stories exploit the extreme time dilation of near-light-speed travel and the associated one-way trip to the future, including Joe Haldeman's *Forever War* and Poul Anderson's *Tau Zero*. (Not to worry: We'll get to relativistic time dilation later in the article.) Yet others, like Andre Norton's

2. That distance is current as I write in January 2011. But Voyager just keeps coasting away, and you can watch the distance grow at http://voyager.jpl.nasa.gov/. (When I looked again in December 2011, the distance had grown to 17.93 billion km.)

3. December 2011 mea culpa: I should have been more specific, fastest relative to what? *Voyager 1* is the fastest object headed away from the sun (a bit faster than sister spacecraft *Voyager 2* and the presently Pluto-bound *New Horizons* probe).

   But not the fastest spacecraft of all. The *Helios A* and *B* probes, long out of service, are in highly elliptical orbits around the sun. At their closest approach, or perihelion, the Helios probes clip along at about 70 km/sec (versus about 17 km/sec for the outbound *Voyager 1*). See http://io9.com/5786083/what-are-the-fastest-spacecrafts-ever-built

4. http://imagine.gsfc.nasa.gov/docs/ask_astro/answers/980317b.html

*The Stars Are Ours* and Vernor Vinge's *A Deepness in the Sky*, put the crews into suspended animation for the long, slow trip.

But other stories demand FTL. Galactic empires (as in Isaac Asimov's Foundation series) and far-flung interstellar trading organizations (like Poul Anderson's Polesotechnic League) become implausible things if the time to reach across the territory exceeds anyone's lifetime. (Hypothetical medical advances might circumvent the lifetime issue, but not the divergence of civilizations increasingly out of sync with one another.) Nor does interstellar conflict seem promising as a story premise if no one in the government will live to hear the outcome of the military actions they order.

### *Starfaring without FTL*

As our *Voyager 1* experience suggests, chemical rockets aren't going to support interstellar travel. Even to loft a satellite into low Earth orbit, at a velocity of ~8 km/sec, burns a lot of fuel. Most fuel serves to carry the fuel to carry the fuel to … carry the payload.

We know how, in principle, to do better. That is, we can conceptualize propulsion technologies that rely upon fuels of a much higher energy density (read: nuclear) and/or in which ships don't carry fuel at all. Examples:

- Solar sailing. Light from a star pushing against a large surface delivers a slow but steady acceleration over a long time. The Japanese Ikaros probe successfully demonstrated the technology in 2010.[5]
- Beam propulsion. Sunlight dims rapidly with distance (a pesky inverse square law is at work), so substitute a high-intensity laser beam. The beam diverges slowly with distance, and so performs better than strictly solar sailing. (Most likely, a space-based, solar-powered, phased array of lasers would be used. Atmosphere would neither dilute the sunlight power-

---

5. See http://www.space.com/missionlaunches/worlds-first-solar-sail-photo-japan-100618.html.

ing such an array nor scatter its beam.)

- Magnetic sailing. The magnetic field projected by one's ship is repelled by the sun's magnetosphere.
- Project Orion.[6] Lob a nuclear device behind your ship. The explosion presses against a pusher plate (remember to put a *big* shock absorber between the plate and crew portions of your ship). Repeat.
- Fusion rocket. Fuse hydrogen into a very hot (read: high speed) helium exhaust.
- Bussard ramscoop. Your fusion rocket projects magnetic fields to sweep up hydrogen fuel from the diffuse interstellar medium. Fuel collection increases with the ship's speed and (the somewhat variable) density of interstellar gas. Onboard fusion fuel is needed for preliminary acceleration and final deceleration.[7]

As a quantitative example, consider Project Daedalus: the British Interplanetary Society concept for sending an unmanned probe to Barnard's Star. Barnard's Star is, after Alpha Centauri, the sun's closest neighbor, "only" 6 LY away.

The Daedalus mission envisions using a fusion rocket. Setting off with fifty thousand tonnes—one tonne, a metric unit, equals 1.1 American standard tons—of helium-3 and deuterium as fusion

---

6. http://en.wikipedia.org/wiki/Project_Orion_%28nuclear_propulsion%29

7. Or a ramscoop might not work at all. Much of the interstellar medium takes the form of hydrogen ions, and a magnetic field moving through a sea of ions experiences drag. You can find arguments on both sides of whether a Bussard ramscoop is practical.

For people in the "won't work" camp, a silver lining: magnetic wings such as a Bussard ramscoop would deploy may offer an excellent way to slow down at the end of an interstellar journey.

fuel, reaching a peak velocity of 12% of light speed, the spacecraft delivers a 450 tonne payload into orbit around its target star after fifty-six years.

### *Why interstellar travel is hard*

No matter the propulsion technology, we are adding energy to an object to impart velocity. From Newtonian mechanics (that is, at velocities too low for relativistic effects to manifest), the kinetic energy (KE) intrinsic to an object of mass *m* and velocity *v* is:

**(1)** $KE = (1/2) * m * v^2$

If we aspire to cross interstellar distances in less than millennia, that velocity-squared factor is problematical. Consider spacecraft A and B, with B moving twice as fast as A. B has *four times* the kinetic energy of A. (Read: B's propulsion system must impart four times the energy.) If spacecraft C is three times as fast as A, C has *nine times* the kinetic energy of A. And so on.

How much kinetic energy is in a one-kilogram mass moving at one-tenth light speed?[8] By equation (1), and expressing light speed in meters (not, as before, kilometers) per second,

$$
\begin{aligned}
KE &= (1/2) * m * v^2 \\
&= (1/2) * 1\ kg * (3 \times 10^8\ m/sec)^2 \\
&= 0.5 * 1 * (9 \times 10^{16}\ kg\ m^2/sec^2) \\
&= 4.5 \times 10^{16}\ joules
\end{aligned}
$$

A joule, more familiarly, is the energy that produces one watt of power for one second. A typical nuclear power plant has a capacity of ~1 gigawatt ($10^9$ watts). So: the kinetic energy of a one-kilogram mass moving at one-tenth light speed equals the energy output of 100

8. Coasting takes no new energy. The challenge is to *accelerate* the object up to its coasting speed. No energy transfer is 100% efficient, so the actual energy expended will always exceed the kinetic energy of the accelerated object.

nuclear power plants running at full capacity for 450,000 seconds (~5.2 days). Not an easy amount of energy to impart to an object …

It's more than 4 LY to the nearest star. A ship and crew will weigh much more than one kilogram. And a ship expends as much energy to stop as to start.

Interstellar travel is challenging.

### *Why relativity makes things really hard*

Albert Einstein's special relativity takes as axiomatic—it does not derive—that the speed of light is constant. Acceptance of SR demanded a revolutionary shift in physicists' worldview: for the speed of light to be always and everywhere the same, space and time must be malleable. But Einstein's insight served to explain the otherwise enigmatic results of the Michelson-Morley experiment.[9]

(Here's a quick overview of that enigma. In algebra texts and Newtonian physics, speeds add. A car northbound at 60 mph and a car southbound at 60 mph converge at 120 mph.

Earth has an orbital velocity of ~30 km/sec. Send light beams in the direction of Earth's orbital motion, and perpendicular, and behind. This time, we do not see velocities adding: all three light beams have the same measured speed. Speed = distance/time, and to make the math work out, time and/or distance must change. Hence, relativity speaks of four-dimensional space-time.)

From the constant-speed-of-light axiom, by fairly straightforward math, flowed consequences that could be—and have been repeatedly—verified.

Special relativity predicts that the faster an object moves, the more massive it becomes. Consider a mass $m$ (measured at rest)

---

9. http://en.wikipedia.org/wiki/Michelson%E2%80%93Morley_experiment

moving at a constant velocity *v* (relative to you).[10] The speed of light—relative to anyone—is denoted by *c*. The relativistic mass *M* for that object is:

**(2)** $M = m / \sqrt{(1 - v^2/c^2)}$

To date, humanity's space ventures have involved $v \ll c$. (*Very* much less. *Voyager 1*, its speed boosted by the gravitational-slingshot effect of its Jupiter close encounter, is tearing away from the sun with a relative speed of . . . less than 18 km/sec. That's about $6 \times 10^{-5}$ of light speed.) In these circumstances, the $v^2/c^2$ term is effectively zero, the square-rooted quantity is effectively 1, and so $M \sim m$. But if we plan to travel between stars within, say, a human lifetime, then v needs to be an appreciable fraction of c. And as (2) informs us, the faster a mass moves, the larger its relativistic mass:

| *v/c* | *M/m* |
|---|---|
| 0.1 | 1.005 |
| 0.2 | 1.021 |
| 0.5 | 1.155 |
| 0.9 | 2.294 |
| 0.99 | 7.089 |
| 0.999 | 22.366 |

With each increment of velocity, the object becomes yet more massive. (That statement is not mere theory: relativistic mass is a factor in the operation of particle accelerators, wherein effects on the orders

10. Situations with an accelerating frame(s) of reference fall—pun intended, if you get it—into the domain of general relativity (GR), and are quite messy mathematically. Exact solutions wouldn't change my basic points.

(A probably apocryphal story has it that in the early days of GR someone told astrophysicist Sir Arthur Eddington that he [Eddington] must be one of only three people in the world who fully understood GR. Eddington is said to have gone silent—not from modesty, but trying to think who the third person could be.)

of thousands are observed.[11]) So: at high speeds, the kinetic energy of a moving object increases rapidly with mass *and* velocity.

And the source of that additional mass? It's converted from (is a transformation of) energy. From the most famous equation in SR (or, possibly the entirety of physics):

**(3)** $E = mc^2$

As $c^2$ is a large number, manifesting even a tiny amount of mass entails a lot of energy.

As v approaches c, the square-rooted quantity in (2) approaches zero—and for any nonzero value of rest mass, M approaches infinity. The energy to accelerate an object (no matter how little its rest mass) to light speed is infinite.

So: no object having mass can reach the speed of light. Photons, the "particles" of light itself, have zero mass. (Neutrinos *do* have mass. That's one reason why the 2011 reports of FTL neutrinos are so puzzling. How could they have been accelerated to that speed?)

### *More complications*

Suppose we build a near-light-speed starship. As it plows through the interstellar medium, its speed converts otherwise harmless gas and dust into highly energetic projectiles: cosmic rays. Our ship needs yet more mass, as shielding. Ionized particles can be deflected by electromagnetic shielding, but not all particles are ionized.

Our ship will also encounter … photons. Just as a fire-truck siren sounds higher pitched—of a shorter wavelength—as we drive toward it, the light from the star we approach is shifted to a shorter wavelength. The shorter the wavelength of light, the higher its frequency and the higher the energy of each photon.

The faster our ship, the bigger the shift. What would otherwise be harmless visible-wavelength light strikes our ship as ultraviolet,

---

11. http://galileoandeinstein.physics.virginia.edu/lectures/mass_increase.html

then (as we go faster) X-rays, then (as we go faster still) as gamma rays. Photons carry no electric charge; like neutral matter, they cannot be deflected by electromagnetic shielding.

Shielding makes our ship yet more massive—and harder to accelerate to high speeds.

### *Tick-tock, tick … tock*

When a spacecraft moves at near-light speeds, more changes—as seen from our Earth frame of reference—than the ship's mass. A clock aboard the spacecraft ticks slower than a clock that is at rest. (Time dilation has been confirmed, too: broadcasts from GPS satellites must correct for this effect.) For a ship of mass $m$ traveling at velocity $v$, a shipboard interval of time $T$ relates to the interval $t$ observed on the ground according to the relationship:

$$(4)\quad T = t / \sqrt{(1 - v^2/c^2)}$$

For $v \ll c$, the square-rooted quantity is effectively one; shipboard and earthbound clocks measure time's passage similarly. For v of ~0.866 c, the denominator in (4) reduces to one-half. The clock ticks once aboard ship for every two ticks on the ground. Put another way, the shipboard clock runs at half the pace of the earthbound clock. For values of v approaching c, the denominator approaches zero; an observer on Earth sees the shipboard clock at a standstill.

Special relativity also predicts that a fast-moving object physically contracts in the direction of motion. Let a ship approach light speed and—you guessed it—its size (in the direction of travel) appears to shrink toward zero.

### *Where SF has gone before*

Not only do many SF authors have great physical intuition, some are practicing physicists. Let's survey the genre for FTL candidates.

The infinite improbability drive (*The Hitchhikers Guide to the Galaxy*, Douglas Adams) taps the Heisenberg Uncertainty Principle. Put simply, that tenet of quantum mechanics (QM) bounds the precision with which one can simultaneously determine an object's location and its momentum (= mass * velocity). With a vanishingly small probability, an object might be *anywhere*. Now merely learn to manipulate probability ...

Okay, not a real FTL-drive candidate, but it *is* really amusing.

(One way in which the uncertainty principle really does manifest itself is in a phenomenon called quantum tunneling, wherein a subatomic particle that—by classical mechanics—lacks the energy to escape its confinement nonetheless does. And so an alpha particle, tightly bound for ages inside a nucleus, suddenly, and randomly, just appears outside. U-238 has a half-life of ~4.5 billion years, so one might be excused for imagining a tiny infinite improbability drive aboard the occasional jail-breaking alpha particle.)

Speaking of the imaginary, consider the inversion engine (from the Skolian Empire series, Catherine Asaro). Equations (1), (2), and (4) become problematical—have divide-by-zero problems—when v equals c. Getting from v<c to v>c there is, intuitively, a moment when v=c. But that intuition presumes velocity values lie along a number line.

Mathematicians offer an alternative to the number line: the number plane (and, doubtless, in some remote, ivied hall, higher dimensionalities), populated with complex numbers.

On a number *plane*, so-called real (i.e., familiar positive and negative) numbers shrink and grow as we move parallel to, say, the X axis. So-called imaginary numbers, both positive and negative, shrink and grow as we move parallel to the y axis. (What's an example of an imaginary number? The square root of negative one, often denoted by the letter i.) A complex number (e.g., -3+2i) on the number plane has real and imaginary components.

Suppose you find a way in which complex numbers apply to velocity. Then you can cross from v<c to v>c in the real-number-velocity component somewhere on the number plane with a *non*zero value of the imaginary-number-velocity component. Goodbye, divide by zero; hello, FTL drive.

Does an imaginary component to velocity make sense? Not in any known way. That said, electrical engineers need complex numbers to describe matters as commonplace as alternating-current electrical circuits. *Something* about complex numbers applies to the real world, so who knows …?[12]

In *The Sins of the Fathers* Stanley Schmidt offers the Rao-Chang drive, by which spaceships jump discontinuously from a slower-than-light to a faster-than-light speed. The analogy subliminally offered to the reader is to quantum tunneling. (*Is* that how the drive circumvents the light-speed barrier? Unclear. In the novel, Chandragupta Rao disclaimed any knowledge of the drive's underlying physical principles, saying that he and his co-inventor stumbled across it.)

That light sets a physical speed limit does not prove that c is a universal, eternal constant. In *A Fire Upon the Deep*, Vernor Vinge divides the galaxy into strata with different physical properties—including separate speed limits. (Earth, alas, lies deep within the Slow Zone.)

Might real physical constants be localized? And, maybe, *not* constant?

Big Bang theory successfully explains many characteristics of the observed universe. Some observations, however, stubbornly refuse to fit the model without an ad hoc workaround called "cosmic inflation." Per inflation theory, in an instant the very young universe expanded by a factor of $10^{28}$ or more. (A *very* ad hoc adjustment. I've seen inflation models with an expansion factor of $10^{60}$.) The light-speed limit governs movement *through* space-time, not *of* space-time, and the early universe spread so quickly that light from some regions will never reach us.

With inflation integral to mainstream cosmological thought, perhaps a starship might (pay no attention to the hand-waving behind the curtain) alter space-time in its vicinity. Within a bubble of altered space-time, local c might be much faster than exterior c. And the ship carries that bubble around with it.

---

12. If you are reading this, Catherine, I suggest you trademark iFTL before Apple does.

(And outside the mainstream? A speed of light that was much faster in the early universe than in our era might explain astronomers' observations without recourse to cosmic inflation.[13])

As yet another possibility, consider that, per quantum-mechanical theory, "empty space" isn't exactly empty. The purest vacuum seethes with "quantum foam": a froth of spontaneously forming—and mutually-destroying—particle/antiparticle pairs.

(How can that be, you ask? Like the indeterminacy bounds governing (and obscuring) a particle's exact position and momentum, the Heisenberg Uncertainty Principle entails an unavoidable fuzziness about a particle's lifespan and energy. And so: particles can be—and apparently are—briefly summoned from nowhere without violating conservation of energy.)

Due to yet other QM strangeness, every particle has an associated wavelength. Now imagine a region physically constrained so as to exclude long wavelengths. What results is the Casimir effect: a measurable force that in a vacuum will push together two parallel, electrically neutral plates.[14]

The region from which some particle-pair types are excluded is said to be energy depleted. The physical properties therein, including light speed, (may) vary from those of a non-energy-depleted region. One more way (perhaps) to wrap a faster-than-normal-c bubble around one's starship.

Next: a recent astrophysical study derived different values for the fine-structure constant from measurements taken looking deep into space along different directions.[15] In one direction the fine-structure

13. See *Faster than the Speed of Light: The Story of a Scientific Speculation*, by João Magueijo. If he's right, maybe we can learn to modify light speed to something friendlier for interstellar travelers.

14. http://en.wikipedia.org/wiki/Casimir_effect

15. http://www.livescience.com/space/fine-structure-constant-varies-space-100915.html

constant (denoted α [alpha]) appeared to decrease with distance. In another direction, α appeared to increase. α relates to the fundamental nature and strength of electromagnetic interactions—and varies inversely with the speed of light. Again, sublight travel in some locales might exceed light-speed travel in our home environs.

However you alter space-time around your starship, the respectable name for the concept is (after physicist Miguel Alcubierre) the Alcubierre warp drive.[16] The *Star Trek* warp drive seems to be of this class.

If (a) a ship in an artificially generated space-time bubble moves through (b) unaltered space-time at what is in *our* frame of reference FTL speeds, while (c) within that bubble the ship never exceeds its local light speed—is that an FTL ship?

### *Warped thinking*

To speed characters across interstellar space, the movie *Stargate* and its spin-off TV series create wormholes between widely separated portals. Some *Star Trek* episodes use naturally occurring wormholes. In Larry Niven and Jerry Pournelle's *The Mote in God's Eye* (in the UK, *The Moat Around Murcheson's Eye*), the Alderson Drive works between naturally occurring interface points.

A wormhole (if any exists) is a topological construct, the analogy being that a worm might get "around" an apple faster by going through it than by crawling along the skin. A better analogy is connecting two dots (think: two stars) far apart on a sheet of paper (think: space-time). Conventional space travel is like tracing a line across the paper, from one dot to another. Wormhole travel has the paper fold/sag until the dots meet.

In short, the wormhole is a cosmic shortcut. Points that are light-years apart as astronomers see them might be—through such a shortcut—quite close. A ship taking the shortcut has an effective speed (= [conventionally measured distance] / [time to cross the wormhole]) that is FTL.

---

16. See: http://en.wikipedia.org/wiki/Alcubierre_Warp_Drive

No one has ever seen a wormhole, but (a) general relativity equations permit wormholes to exist in sufficiently warped regions of space-time and (b) GR describes gravity as the warping of space-time caused by the presence of mass.

What would it be like to cross such a wormhole? Like grazing the event horizon of a black hole—but at just such a trajectory that rather than be crushed, you shoot through the shortcut to exit a white hole.

A white hole, as you might expect, is a hypothetical egress from a black hole. Black holes, by definition, can't be seen directly (although the radiation from in-falling matter offers a very good indication). White holes, on the other hand, should be directly visible—if they exist.

Perhaps no white hole has been seen because GR solutions for many wormholes are unstable. The math predicts that static wormholes will collapse faster than anything could cross through. That said, other solutions, involving a massive spinning ring, perhaps made of exotic matter exhibiting negative energy density, *are* mathematically stable. (Not that exotic matter with the required properties has been shown to exist. Regardless, it may be possible to create an energy-depleted space via the Casimir effect to stabilize a wormhole.)

Here's a NASA description[17] of one way to build a wormhole:

> First, collect a whole bunch of super-dense matter, such as matter from a neutron star. *How much?*—well, enough to construct a ring the size of the Earth's orbit around the Sun. Then build another ring where you want the other end of your wormhole. Next, just charge 'em up to some incredible voltage, and spin them up to near the speed of light—both of them.

Perhaps no one has seen wormholes because naturally occurring

17. From http://www.nasa.gov/centers/glenn/technology/warp/ideachev.html.

ones are too small and short-lived. Likely *really* tiny black/white hole pairs are constantly manifesting out of—and disappearing back into—the quantum foam. If so, the trick will be to spot and separate such infinitesimally small objects before they mutually destruct, grow them to useful size, prevent them from collapsing, position them to form a wormhole, and move one or both ends to a useful location.

Wormhole engineers (if their subject matter proves real) have their work cut out for them.

### December 2011 update: FTL neutrinos

On September 22, 2011—after six months spent looking in vain for errors in their results—a team of particle physicists affiliated with the European Organization for Nuclear Research (best known as CERN, its French acronym), announced they had measured neutrinos traveling twenty parts per million *faster* than light in a vacuum.[18]

The physics community and the popular press have been abuzz ever since.

(Neutrinos—Italian for "little neutral ones"—are subatomic particles. They mass so little that physicists were unsure for many years whether neutrinos have any mass at all. Among their unusual attributes, neutrinos spontaneously change among several types. Neutrinos by the billions pass through your body every second without effect—that's how weakly they interact with more familiar subatomic particles.)

On the one hand: as I write, no one has found an error in the experiment. The group at CERN has even rerun their experiment with one minor change, duplicating their original result while eliminating one variable (among many) that might have led erroneously to the surprising conclusion.

### On the other hand ...

No one has independently confirmed the result, either. (To be complete, the CERN team is the *second* to report FTL neutrinos—

18. http://www.guardian.co.uk/science/2011/sep/22/faster-than-light-particles-neutrinos?INTCMP=ILCNETTXT3487

but the prior instance, reported in 2007 by a team affiliated with Fermilab, involved too few neutrinos to be statistically significant. Fermilab is among the research organizations eager to duplicate or refute the CERN results.[19])

Some skepticism about the CERN report concerns the complicated mechanics of the experiment. Speed equals distance divided by time. The experimenters must know *exactly* both the flight time of the neutrinos and the distance between (a) the particle accelerator at CERN in suburban Geneva, Switzerland, and (b) the neutrino detector in Gran Sasso National Laboratory in central Italy. How exactly? To nanoseconds and nanometers. Both measurements relied upon the Global Positioning System—thus involving, among other factors, relativistic corrections for the orbital motion of the GPS satellites. And then there is the notorious difficulty in detecting any neutrino.

It's not like timing the hundred-yard dash with eyeballs and a stopwatch.

Meanwhile, some theorists argue that FTL neutrinos must lose energy because of Cherenkov radiation: a subtle effect observed in many experiments involving speeding subatomic particles. The neutrinos in the controversial CERN experiment were *not* seen to radiate and lose energy in the expected manner.[20]

Astrophysicists refer to 1987A, a supernova event observed in a galaxy that's about 160 thousand light-years distant from Earth. Neutrinos and then light from 1987A reached Earth mere hours apart. (That short gap has been attributed to differences in how neutrinos and photons interact with matter. Neutrinos scarcely interact; they would stream unimpeded from an exploding star. In contrast, photons would be absorbed and re-emitted over and over

19. http://idealab.talkingpointsmemo.com/2011/09/was-einstein-wrong-us-accelerator-lab-to-test-cerns-results.php

20. http://www.sciencenews.org/view/generic/id/335309/title/Critics_take_aim_at_fast_neutrinos_

on their route to the star's surface.) If neutrinos outraced light to the extent reported by the group at CERN, 1987A's neutrino burst would have arrived four *years* in advance of the light.[21]

If physicists eventually confirm faster-than-light neutrinos—and those who would refute Einstein have yet to succeed—two things are certain. Much of the last century's physics will have been stood upon its metaphorical head. And I'll have exciting material for a future essay.

Now let's return to the essay as I first wrote it.

### *Beyond Relativity*

To my knowledge, no reproducible observation contradicts Einstein's relativity theories. But Isaac Newton's mechanics explained the observations of *his* time and for centuries after. Einstein's solutions may also prove to be merely an approximation sufficient for known phenomena—and fail to explain something observed tomorrow.

In support of that possibility, we *know* relativity theory is incomplete.

General relativity (GR) considers the overall structure of space-time, in which gravity is a manifestation of the curvature of space-time caused by the presence of mass. GR deals well with matters at a large scale. Crucially, the GR equations assume space-time is continuous.

Quantum mechanics, another long-lived—and very complex—physical theory deals with matters at a very small scale. In QM, matter and energy come in discrete, indivisible chunks called quanta. The QM worldview is inherently *dis*continuous.

Among the grand challenges of modern physics is a Theory of Everything reconciling GR and QM. (That can't come soon enough: It will take unifying quantum scale events with gravitational/space-time effects to fully characterize the Big Bang. Or black holes.) It's expected that reconciliation will add quantization to our understanding of space-time. If space-time consists of indivisible quanta,

---

21. http://blogs.discovermagazine.com/badastronomy/2011/09/22/faster-than-light-travel-discovered-slow-down-folks/

however tiny, divide-by-zero limits may disappear from revised relativity equations.

String theory is (at least in terms of the number of physicists working in the specialty) the leading approach toward reconciling GR and QM. *Extremely* briefly, where QM sees the fundamental building blocks of matter, such as electrons and quarks, as dimensionless points, string theory sees those building blocks as vibrations of one-dimensional strings in an N-dimensional space-time. String theory isn't *a* theory; rather, it is an umbrella term for many families of mathematical descriptions, each variant offering its own set of dimensions and mathematical rules. As I write, the favored number of dimensions among string theorists is eleven.

Eleven dimensions? Einstein's space-time has four. You and I don't experience the extra dimensions—as the theorists would have it—because the extra dimensions curl up at scales too small for us to see. Alternatively, our 4-D space-time and the other dimensions have a very small intersection (just as an ideal sphere resting upon an ideal plane touch at only one point.)

(December 2011 aside: One theory to explain CERN's apparently FTL neutrinos is that they took a shortcut through a hidden dimension.)

### Hyperdrive

If you can't go faster than light in space, then why not go under or over space? That's the premise of (one name among many) hyperdrive.

Jack Campbell's Lost Fleet series uses FTL through "jump space" between rare naturally occurring access points *and* a faster FTL between manmade portals of a hypernet. In constructing a military SF plot, one wants to provide targets; access points to the stars fill the bill. Niven's Known Space (in which I've sometimes coauthored) offers a hyperdrive. Hyperdrive works most anywhere outside the depths of gravity wells.[22] Vinge's hyperdrive works outside the galactic Slow Zone, with pace of FTL varying within and between galactic zones.

Physicists have yet to dub anything hyperspace, but some do—see earlier discussion of string theory—speak of hidden dimensions. Sounds like a potential hyperspace to me.

In the variant of string theory called M-Theory, M for membrane, our familiar universe is a 4–D island adrift in a higher-dimensional space. We're wrapped, at least conceptually, in a membrane that separates us from those other dimensions. (If this theory is correct, lower-dimensional universes like ours can collide with one another. I suspect that that would be a bad thing.) Physical laws beyond the membrane may differ dramatically from laws within the membrane.

"Beyond the membrane" sure sounds like SF's hyperspace.

As a matter of semantics, one could again argue whether travel outside our universe at a speed faster than c constitutes a violation of relativity inside our universe.

### *FTL communications*

Some stories offer FTL communications. In the Ender universe of Orson Scott Card, ships travel at sublight speeds but can communicate instantaneously via an "ansible."[23] Niven's hyperdrive has a finite speed, but his hyperwave radio (where it works, which is again

---

22. Known Space stories use the term *gravitational singularity* for a region whose gravity interferes with a hyperdrive. Being near any star, or even a decent-sized free-flying planet, is more than sufficient to discombobulate a hyperdrive shunt. *Gravity well* is the more general term—any mass warps space-time at least a little—whereas a singularity commonly refers to a point of infinite density (such as a black hole).

23. Per Wikipedia (http://en.wikipedia.org/wiki/Ansible), Ursula K. LeGuin invented the ansible in *Rocannon's World*, and Card (and others) adopted it. I simply don't know.

outside deep gravity wells) is instantaneous.

Are there ways to imagine FTL, or even instantaneous, communications? Besides, at this point in the article, the obvious: light beams and radio waves sent through any channel, such as a wormhole, used for FTL travel.

Maybe.

Among the odder attributes (a high hurdle) of quantum mechanics is quantum entanglement. *Very* briefly, consider two like particles—say, two electrons—that have become associated. Their states differ only in one quantum parameter, such as the orientation of their spins, of which only two values can be measured. (I choose my words very carefully here. Stay tuned.) QM does not allow two particles to have identical parameters; anytime the spin state of one particle is measured in either of its permissible two values, the spin state of the second particle takes the other permissible value. Instantaneously. Even if someone had previously relocated one of the linked pair halfway across the universe.

Einstein called this connection between paired particles "spooky action at a distance" and understood it as a crack in the foundations of then-young QM theory. Nature begged to differ; starting in the 1980s, instantaneous state changes through quantum entanglement have been demonstrated across separations of miles.

So can one use quantum entanglement for FTL communications? Due to other quantum weirdness, no. Above, I carefully referred to two permissible measured values for the electron spins. Between measurements, the spin state of any particle is … indeterminate. (Think: Schrödinger's cat.) Through measurement I can instantaneously force a discrete spin state upon the near particle—and hence also upon the remote particle—but I can't know what state will be evoked where. I might (for example) only be confirming the probabilistic value the remote particle had before the determination.

When no information is sent, there is no communication.[24]

---

24. People drive themselves crazy looking for a way to encode information into this instantaneous signal. The consensus remains that no information can be sent.

    For more about quantum entanglement (including applications that *do* work, like quantum cryptography), see "The Strange World of Quantum Entanglement," by Paul Comstock (http://calitreview.com/51).

While it's common to say that special relativity prevents anything from outpacing light, the more precise statement is that *information*—in a physical object or otherwise—cannot travel faster than light. A "signal" from which no data can be extracted *can* exceed the speed of light.

As another possible FTL comm mechanism, consider the tachyon. A tachyon, from the Greek *tachys*, or swift, is a hypothetical fundamental particle that *must* travel faster than light. (*Very* hypothetical. Tachyons do not have a home in the Standard Model of particle physics. While the Standard Model is not 100% settled science—it predicts a particle, yet to be observed, called a Higgs boson to explain why matter has mass—the model has avoided serious challenge for decades.[25]) Tachyons never cross the v=c threshold, avoiding the divide-by-zero problems in the relativity equations. No tachyon has ever been observed (and I'm going to gloss over if/how an experiment could detect one), nor do tachyons, strictly speaking, violate any known laws of physics.

But if tachyons do exist, they raise new vexing issues.

### *For the last time (he says, ironically), back to special relativity*

We've seen that mass and time dilation derive from special relativity's central tenet of light as the universal speed limit. Here's another derivable consequence of that tenet: an object or signal traveling faster than light would be (to the rest of us) moving backward in time. As SF readers well know, time travel to the past creates paradoxes.

Of which perhaps the best known is the Grandfather Paradox. If I travel back in time and kill my own grandfather, how did I come to exist? But if I never existed, then clearly I didn't kill my grandfather. So I do exist, and can travel backward in time. So …

Great authorial fun, to be sure. See "Grandpa?" next up in this collection.

---

25. Per a 2011 update to the earlier essay about RFIDs, the Higgs boson may finally have been glimpsed, although not yet at the level of statistical significance scientists generally require.

To my knowledge, no one has ever seen a time traveler, so maybe the universe defends itself against such paradoxes. Maybe the universe is its own time police.

To abandon the precedence of cause over effect would strike at the very foundations of our sense of reality; most physicists are loath to give credence to tachyons until they see one.

Many fictional uses of FTL choose to ignore the backward-in-time implication. Among the stories that embrace the implication are James P. Hogan's *Thrice Upon A Time*, Gregory Benford's *Timescape,* and Roger MacBride Allen's *The Depths of Time*.

(Do you see now why the previous caption was ironic?)

### Wrapping up

Are FTL travel and FTL communications merely SF tropes? Not always. Plenty of science at the cutting edge—and we exhausted our page budget well before we exhausted the topic—deals with concepts that might create FTL technology.

Doubtless with implications that are challenging to anticipate.

That's okay. Foreseeing consequences is among the higher callings of science fiction.

### To read further

*Interstellar Travel and Multi-Generational Space Ships*, edited by Yoji Kondo, Frederick C. Bruhweiler, John Moore, and Charles Sheffield.

*Hiding in the Mirror: The quest for alternate realities, from Plato to String Theory (by way of Alice in Wonderland, Einstein, and The Twilight Zone)*, by Lawrence M. Krause.

"NASA Goes FTL Part 1: Wormhole Physics" and "NASA Goes FTL—Part 2: Cracks in Nature's FTL Armor," Alternate View columns in *Analog* by John G. Cramer; respectively the Mid-December 1994 (online at http://www.npl.washington.edu/AV/altvw69.html and February 1995 (http://www.npl.washington.edu/AV/altvw70.html) issues.

# *Grandpa?*

The lecture hall was pleasantly warm. Behind Prof. Thaddeus Fitch, busily writing on the chalkboard, pencils scratched earnestly in spiral notebooks, fluorescent lights hummed, feet shuffled. A Beach Boys tune wafted in through open windows from the quad.

"And so," he continued, "travel backwards in time would violate causality, and hence appears to be impossible." He turned to face the class. "The problem is most commonly illustrated with the 'Grandfather Paradox.'

"Imagine that I had the technology with which to visit my grandfather in his youth. Once there, what is to stop me from killing him before he'd had the opportunity to reproduce? But if I did succeed, who was it who had traveled backward …"

A collective gasp interrupted his lecture. As Thaddeus realized that the students were all staring at his chest, he glanced downward to see a red dot glowing on his white shirt. The professor next noticed, in the back row of the hall, a young man who seemed to be aiming some sort of pistol at him. The would-be sniper's hand seemed to shake in unison with the wobbling of the red spot.

The young man was not enrolled in the Introduction to Physics class, yet he looked somehow familiar to Thaddeus. Perhaps it was the angle at which he cocked his head, or the shock of bright red hair, or the piercing green eyes. In a rush of *presque vu* Thaddeus recognized a version of himself.

The modernistic handgun with its (laser?) sight, the anachronistic clothes, the look of fanaticism … it suddenly all clicked. "My grandson, I presume."

His visitor nodded.

"It can't be done. *Something* must cause such an attempt to fail."

His spine tingling despite his heartfelt conviction, Thaddeus's eyes resumed their standard lecture-hall sweep. "Class, as I was saying …"

Two loud shots rang out almost as one. The noises released the students from their paralyzed shock. As they wrestled his assailant to the floor, Thaddeus realized that the first shot had been the backfire of a passing vehicle, and that the unexpected noise had startled his grandson enough to throw off his aim. A hole in the chalkboard showed that causality had needed only inches of deviation to protect itself.

A soft "pop" emerged from the pile-up of students who had tackled the time traveler. The pile collapsed, as if its central support had vanished. As confused students gradually untangled themselves, it became clear that Thaddeus's grandson had, in fact, disappeared.

◆❖◆

Thaddeus's early afternoon lecture started little different from his morning session. The hall was perhaps slightly warmer; the music from the quad was now by the Everly Brothers. The fluorescent hum was occasionally submerged by the buzz of students discussing the morning's excitement.

"Imagine that I had the technology with which to visit my grandfather in his youth. Once there, what is to stop me from killing him before he'd had the opportunity to reproduce? But if I did succeed, who was it who had traveled backward …"

A large pigeon chose that moment to fly through an open window into the hall. Its stately path abruptly veered downward, and it thumped, flapping feebly, to the auditorium floor. Thaddeus had just noticed what appeared to be the handle of a large throwing knife when a mob of enraged students rushed the returned, red-haired attacker.

Once more, the time traveler disappeared from the pile of bodies to the accompaniment of a pop.

◆❖◆

Word of mouth had filled Thaddeus's third lecture of the day to overflowing. The crowd warmed the hall to an uncomfortable degree; expectant muttering masked any music that might have been playing on the quad.

"Imagine that I had the technology with which to visit my grandfather in his youth. Once there, what is to stop me from killing him before he'd had the opportunity to reproduce? But if I did succeed, who was it who had traveled backward …"

Thaddeus's now all too familiar look-alike appeared in the aisle behind the last row of lecture seats. Moments later, an ovoid object came hurtling toward the professor.

Mid arc, the grenade disappeared with a pop. His grandson vanished with a second pop.

◆❖◆

Order eventually returned to the lecture hall. Students took their seats. All eyes turned questioningly to the professor.

"I believe that we've seen the last of our troublesome visitor," said Thaddeus.

He paused. He shrugged. He smiled.

"I've decided that I'll never have children."

# *Great Minds*

"It's very much as I expected," the intruder said without preamble.

Entering my cozy den, I had encountered him seated in my massive leather wingchair, shoes up on my broad mahogany desk, savoring one of my Cuban cigars. A snifter of brandy rested on the leather blotter, within his easy reach. The aroma was Napoleonic.

As I was unsurprised to find him. "Please, don't get up."

"You're very gracious." He grinned. The smile was world-famous: toothy, and slightly off-kilter. I saw it every morning in the mirror. Not that there weren't differences. There always were: in haircut, clothing style, glasses instead of contacts, whatever. I found his sideburns curiously short. "I mean considering."

Considering, as we both knew, he was here to take my life. Leather squeaked as his feet swung from the desk and he straightened his posture. Getting down to business. "The greatest minds of the millennium could not reach a common understanding what the math meant." Meaning: He couldn't have been expected to figure it out.

He was a whiner, a self-justifier—for which I was grateful. That character flaw was the only reason I was still here. He was also wrong. Proof by counterexample: *I* had decided I would solve the puzzle. Eventually, he had made the same choice. And, in our own times, in our respective ways, each of us had been successful.

His over-rehearsed rationalization tumbled out. "Bohr, Heisenberg, Einstein, Pauli, von Neumann, Schrödinger, Planck … them and more. Giants. You know the list. They never agreed on the physical significance of the math. Who was I to hope to understand the reality underlying the mathematical formalism of quantum mechanics?"

Meaning: *He* lost hope, and somehow it became justifiable that I should pay the piper.

"And so for a long time, I gave up. I denied the problem. My career went another way." He paused for a sip. "But for years, for *decades*, I could not help but wonder. Every day, billions of transistors demonstrated some underlying truth to the theory. Quantum mechanics describes something. I had to know what."

His non-smoking hand, when not busy with the consumption of my best brandy, darted from time to time to pat something unseen in his coat pocket. It seemed to give him confidence.

"And so you returned to physics." I had never left it.

He admired the many plaques and photos gracing the darkly paneled walls of the room. "And so I realized, I *decided*, what you had much earlier. The Copenhagen Interpretation—that certain physical specifics go beyond being unmeasurable, that to even inquire about them represented a misunderstanding of the physical universe—was, if true, an explanation inherently unprovable.

"What *was* provable, if true, was another explanation altogether: the Many Worlds Interpretation. If I could detect other universes, show that events happened in all possible ways, not just in whatever random way 'the wave function collapsed' without cause or explanation in ours, the great QM debate would be resolved. But among the myriads of myriads, for which other universe would I aim? And what evidence of that other place would be unambiguous?"

His nervous pocket-patting was growing more frequent. If my suspicions about the device in that pocket were correct—and who better than I to understand my visitor's thinking—I did not have much time. "And then you realized … if MWI were true, there must be other universes in which another you"—such as *me*—"had stayed the course." My eyes followed his to the Nobel Prize certificate and medallion in their softly illuminated, velvet-lined display case.

Because you got greedy. You saw you need not settle for fame beginning at age fifty-five—my present age, hence your own. You could do better. Much better. By switching places, you could seize the fruits of fame from another you who had proven the MWI years earlier.

*Do you think you are the first* me *to have had that realization?*

Below his line of sight, I clicked my heels twice. The radio beacon thus triggered activated the mechanism hidden within my/his chair.

◆❖◆

There are universes without number. Among the myriads is one where a different quantum outcome was enough to change the career of an unknown microbe. Newton died there in the great plague of 1665, at age twenty-three. The development of physics was, as a result, greatly impeded. Onto that parallel, low-tech plane of existence now materialized a new occupant.

I am not a cruel man. I sincerely hope my recent visitor—and the dozen earlier versions of me—enjoy their opportunity to make real advances in physics.

# *No GUTs, No Glory*

I have a theory about baseball caps and intelligence. Grown men who wear baseball caps move their lips when they read. Grown men who wear baseball caps backwards *can't* read. Okay, I'll make exceptions for actual pro ball players.

And then, in a league of his own, we have Alistair Winkler. The first words I heard from Alistair's lips, plaintively spoken, were, "Have you seen my baseball cap? I can't find it."

Appearances can be deceiving.

◆❖◆

I had just emerged from my office, too bored for the usual charade. You know: where my secretary keeps a client waiting to show how little I need the business. Right.

Winkler was looking all around, befuddled. He was comfortably middle-aged, average in height and spare-tire size. His eyes, behind Coke-bottle-bottom lenses, were muddy brown. It was windy today; my new client hadn't bothered to comb his sparse hair since coming inside—maybe not since the Carter administration. For sure, that was when he'd bought his tweed sport coat. Lapels don't lie.

"Mr. Winkler?" I said. "Come this way, please. My secretary will be happy to look for your cap." It might even have been true. Phyllis was as bored as I.

My client clutched a shoebox that leaked receipts. If only he had misplaced *that*. He mumbled something as I settled him at the conference table.

"What's that?" I asked.

"Doctor Winkler."

"Sorry." Appearances can be deceiving, I thought, reaching for his box. "When you made your appointment, did Phyllis tell you to bring your last three returns?"

"They're in the box."

Meaning they were, at best, folded. Based on the crumbled receipts, I'd have bet on finding the returns wadded up. Too bad I didn't have any money; I'd have won. Sliding the old 1040s back and forth across a table edge to flatten them, I asked. "So what's your field?"

Winkler mumbled again, something about guts. That was less disturbing than his old returns, which, it appeared, he had prepared himself. He sure wrote like a doctor.

"Um," I responded insightfully. "I don't understand medical specialties. Does that make you a GI doc?" As in: It may be shit to you, but it's my bread and butter.

Winkler's hearty laugh surprised me. "I'm a Ph.D. in physics, not a medical doctor. GUT stands for grand unified theory. You're familiar, of course, with the work aimed at unifying the gravitational, electroweak, and strong nuclear forces."

Electro-who? I hid behind an old tax form, where an odd entry caught my eye. "What's this $407,000 in 'other income' from two years ago?"

Winkler scratched his head, then brightened. "Oh, that must be my share of the Nobel."

I did say appearances can be deceiving, didn't I?

◆❖◆

We went through Winkler's old returns, me tsk-tsking at the many missed deductions. Just what I could recover with a corrected last-year return would more than pay my fees.

According to Alistair, Einstein once said the hardest thing in the world to understand is the income tax. Apparently Einstein had had problems with his grand unified theory, too. I think my client had taken on both GUTs and his returns as a dare.

Alistair read physics journals with one eye as we worked. I'd have needed both eyes, absolute silence, and a brain transplant.

Still, I could read the cover. Okay, a handful of words on the cover. "Black holes?" I commented, making conversation. "Pretty heavy stuff. I bet you could even explain where all the mortgage money is disappearing."

Not billions. Hundreds of billions. Even the T word was beginning to crop up in the news. I'm an accountant, and still these were numbers beyond meaning. Down a black hole? Sure. Why not.

But look who I was talking to: a man who hadn't claimed a personal exemption for his youngest child. Maybe that was too much to ask. Alistair, Jr., was only four. The boy hadn't had long to make an impression.

Winkler smiled uncertainly at my little joke, then frowned. His eyes glazed over. I mean, I knew by then that he marched to a different drummer—but suddenly it seemed like the drummer resided on Pluto. Alistair's pen started flying over a scrap of paper, which happened to be the back of a receipt. He shushed me when I cleared my throat.

As abruptly as Winkler had left planet Earth, he returned. "Why yes," Alistair said, "I *do* have a theory about the mortgage mess."

◆❖◆

They make the best piña coladas here in the islands. It must be the fresh coconut milk. Or the lack of extradition. Now, where was I?

Ah, yes. Straight from my chat with Alistair, I began investigating new office space. I sold my car and pawned my office furniture to do it, but—just barely—I signed a lease on a small office only two floors above the main branch of Great Big First American Federal Trust Savings Bank. That isn't exactly the name. After so many mergers, no bank's name made sense. The bank with the cute cartoon eagle in their ads.

Name notwithstanding, the depositors had lost their trust and feared for their savings. The bank was hemorrhaging cash. I knew the feeling, but that wasn't why I wanted this lease.

To the horror of my new landlord, I hunkered down in my unfurnished office. I had barely scraped together enough money to move in. This had to work—and *fast*.

The universe failed to share my sense of urgency. I hate that in a universe.

Days later, dispirited, I needed reassurance. Or clarification. Or electroshock. I arranged to visit Alistair at his house to discuss his returns. Even *he* might have noticed the barrenness of my office. A dead fern would.

Casually, in the middle of a discussion about his itemizable deductions, I asked, "So everything is quantum mechanical? I'm not sure what that means."

He tugged pensively at his hair, and the resemblance to Einstein increased a bit. "Okay, try this. To you and me, matter looks like it's made of stuff—particles, but matter also has a wave nature.

"A particle's position is described by a probability function that resembles the mathematical description of a water wave. The particle isn't at any one location with certainty; instead it has *some* probability of being anywhere. Until you measure it. Then it has to be someplace, of course."

Those two abused words again: of course. Did Winkler ever say anything to which the phrase applied? I nodded.

"Naturally the probability is highest that a particle will be where you expect it—that is, at the crest of the wave. Still, there is a probability that the particle will turn up elsewhere."

Like receipts escaping the shoebox, I thought. Alistair swore he had had lots more receipts.

Alistair brightened as an example occurred to him. "The particles that comprise an atomic nucleus lack the energy to escape from the forces that bind them. Only the particles don't know that."

"Huh," I contributed. So much for clarification. Or reassurance.

Alistair gestured grandly. "Think of a bullet shot into the air. It's too slow to overcome Earth's gravity, so it falls back. In the same way, there are forces that won't let a particle escape from a nucleus. Still ..."

We were circling around to our conversation of the other day, if not by a route as mundane as a circle. Maybe that was because

Winkler grooved on tiny, wrapped-up-on-themselves, invisibly small dimensions. Seven or eight of them? Don't ask me.

"Still?" I prompted hopefully.

Alistair nodded. "There is a low but real probability that the particle exists outside the nucleus. That's why, occasionally, a particle simply *appears* far away. The farther away, the less likely, of course. Given enough atoms, and enough time, things *will* escape by quantum tunneling. Not that there are tunnels, of course."

Of course, *twice*.

Alistair didn't see me flinch. He was on a roll. "That's radiation. That's how a plutonium atom turns spontaneously into uranium. Its nucleus emits an alpha particle."

"Really?" I asked. Double-declining-balance depreciation, I got, and the alternative minimum tax. But this?

"Oh, yes. That's why there is virtually no naturally occurring plutonium on Earth. Over the eons, almost all of it has decayed into uranium."

"And *everything* behaves quantum mechanically," I said, tapping a Schedule B with my pencil, not much liking the skepticism in my voice. My car hadn't been much, but I missed it. What the *hell* had I done? "Stuff shows up in places it can't get."

"If it's made of atoms, it does," Alistair said firmly. "It's all only a matter of probability. Trust me on this."

"I do," I lied.

But soon I would believe. My office, when I returned that evening, thinking about measuring positions and about tunnels that weren't—

Brimmed with cash. Made of atoms.

The next day, the feds took over Great Big First American Federal Trust Savings Bank.

◆❖◆

I'm as human as the next guy, so, yeah, it occasionally bothers me that I don't precisely understand where my money came from. Precision is overrated.

Let me tell you a story.

This high-school class is holding its twenty-fifth reunion. Naturally everyone is checking out everyone else. There's this one guy who no one recognizes: the guy who drove up in the new Jaguar, wearing the Italian suit. The ballroom buzzes with speculation until the class president works up his nerve to ask. "I'm sorry, my friend," Prez says, "but I can't place you. Are you sure you're at the right reunion?"

The mystery guest smiles condescendingly. "To be honest, I dropped out before graduation. Maybe I should leave."

Prez takes another look at that two-thousand-dollar suit and starts sucking up. Mr. Most Likely To Succeed hadn't. "No, no. Anyone who attended Jefferson is welcome. Now what's your name again?"

To make a long story short—if it's not too late—the rich guy no one can place was the class dummy, and he's made a fortune in the restaurant business. The class president swallows hard and asks, "How did you do it? I've heard profit margins are very thin at restaurants."

"Oh they are," agrees the dummy. "I only make ten percent."

Prez is incredulous, remembering the Jag. "Ten percent?"

"That's right. I buy a steak for five dollars and sell it for fifty dollars."

*I* understand the ten percent that I need to, too.

# *Inside the Box*

The lecture hall was pleasantly warm. Behind Thaddeus Fitch, busily writing on the chalkboard, pencils scratched earnestly in spiral notebooks, fluorescent lights hummed, and feet shuffled. A Beach Boys tune wafted in through open windows from the quad.

Or so, in any case, the professor imagined the lecture hall. Chittering, muttering students squirming in their seats this morning drowned out the customary sounds. Or what he thought he remembered to be the customary sounds …

Chalk squeaked as Thaddeus, with more energy than artistry, began sketching a stick-figure quadruped. "I'll explain this cat momentarily, class." Shrödinger's thought-experiment cat. Today's Introduction to Physics lecture introduced the counterintuitive topic of quantum mechanics. "Recall from your reading that the behavior of atoms and their constituent parts cannot be fully described by such conventional characteristics as position and momentum. More precisely, how we think about those descriptive terms must change." He continued drawing as he spoke, the cube in which he was attempting to enclose the cat somewhat out of perspective. He winced as the chalk snapped, its tip caught by the hole that should not be there. Should it?

"In classical physics, we can, with sufficient care and expense, measure to arbitrary precision the position and momentum of any particle. At sufficiently tiny scales, however, nature does not behave as we expect. Instead, in those infinitesimal domains, we discover that certain parameters exhibit heretofore imperceptible granularity or lumpiness—what physicists call quantization. Further, we cannot measure at quantum scales without influencing whatever is being measured. The math is inappropriate for"—beyond—"this class, but a consequence of quantization is that we cannot have absolute knowledge of subatomic particles."

His crude diagram complete, Thaddeus pivoted to face the packed auditorium. "If we know an electron's position quite exactly, we can know little about its momentum. If we know its momentum, we can tell little about where it is. We are reduced to probabilistic descriptions of where the particle may be, and where it may be going." Doggedly, he ignored the arm waving from the second tier of seats. "But can this uncertainty manifest itself in the macroscopic world *we* experience? That is what Erwin Shrödinger set out to consider—"

"Professor." Young Mr. McDowell's tone, although respectful, was quite insistent. The sophomore stood to emphasize his seriousness. "We—the class, that is—we feel we should discuss yesterday's events."

A flood of … memories? … displaced whatever the student said next. A near-miss handgun attack. A flung knife by chance impaling a pigeon inopportunely availing itself of the open window. A hurtling hand grenade vanished in mid arc.

Thaddeus shook himself by the mental lapels. *Nonsense.* Pointing at the board, he continued. "Returning to today's subject, Dr. Shrödinger devised a thought experiment to illustrate quantum uncertainty. My cartoon reveals the inside of the box, but imagine that its walls are quite opaque, quite impenetrable." Beside the stick-figure cat, he drew a tiny square. "This mechanism contains a bit of radioactive material. Detection of a single radioactive decay," and he tapped the board once with his chalk stump, "releases poisonous gas."

He was explaining a decay event as a particle's spontaneous emission from an atomic nucleus—a manifestation of positional uncertainty—when murmurs of protestation stopped him. Hairs rose on the nape of his neck. In the otherwise jammed hall, one cluster of seats remained unoccupied. It was where *something* had happened.

Only it couldn't have.

Mr. McDowell was still, or once again, on his feet. He followed

Thaddeus's gaze to the empty few chairs. "We don't understand about him either, sir. The … intruder."

Heads nodded. Voices rang out in agreement. A hundred pairs of eyes beseeched Thaddeus. He relented. "My unborn grandson, you mean. It's impossible, you know."

"But professor—"

With outstretched arm and firm voice, Thaddeus interrupted. "You know what you saw, you were going to insist. What you, and your colleagues in later sections of the class, all saw. Or what, rather, you've now *convinced* yourselves you saw, after repeated retellings of the tale." He lowered his arm and voice. "Surely there is a simpler explanation than the impossible.

"A time-travel lecture, illustrated with the grandfather paradox, in a hot, stuffy classroom. A passing car backfires. A guest audits the lecture, someone with red hair like mine. Thrill-seeking students attend later sections of the lecture, and their rumor-fed expectations stoke our own fevered imaginations."

Thaddeus took a deep breath. "What I, too, admit to remembering did not happen. It *cannot* have happened. This can only have been an instance of mass hysteria."

"Like UFO sightings," someone called out.

"Or the Salem witch trials," Thaddeus agreed. Better a moment of soft-headed gullibility than to deny causality. Not that he cared for either of his options …

Young McDowell persisted. "Professor, the blackboard has a *bullet hole*. And how do you explain that the attacks stopped? They ended—you ended them—when you announced you would never have children."

Thaddeus braced himself against his lectern. "A hole was surely in the board all along, unnoticed until the suggestive backfire. And our visitor likely vanished by no more mysterious a means than"—and he gestured to the rear of the auditorium—"that rear exit door." Still, his memory insisted his doppelgänger had disappeared—to the future?—from beneath a pile-up of angry students. "Would *you* choose to re-experience our welcome?"

That drew nervous giggles.

"Ladies and gentlemen, yesterday we spoke about cause and effect. Now you claim that my grandson traveled through time to kill me, and that I defeated his attack by my declaration I would have no children.

"If so, no grandson ever traveled back to cause my decision. Will I still make that decision? Might I now *have* children? Doubts blossomed on their faces, and he hammered the figurative nail into the metaphorical coffin. "How, if I halted the attacks by deciding never to have children, can *you* remember my grandson?"

Whispering stopped as Thaddeus rapped the oaken lectern. "Back to Schrödinger's cat. Has an electron, its exact position uncertain, chanced to manifest itself outside an atomic nucleus? That is, has a radioactive decay occurred to cause release of the poisonous gas? Remember, we cannot see inside the box. Class?"

Confusion returned, but of a more academic nature than the controversy just concluded. (*Concluded*, mocked some corner of Thaddeus's thoughts, *or simply set aside?*)

"A show of hands, please. Who thinks the cat is alive?" A few hands rose tentatively. "And who thinks the cat is dead?" More hands. "Not everyone expressed an opinion. Do the rest of you imagine it's a vampire cat—the undead?"

The chuckle was overlong and overloud. He wasn't the only one still on edge. "In the closed system of the sealed box, we cannot know the cat's status. Neither living nor dead *is* the correct answer—at least by the formalism of quantum mechanics. There is only probability until the box is opened and an outcome observed. Until then, all possible outcomes are said by physicists to be in superposition."

A familiar arm waggled. Thaddeus managed not to sigh. "Yes, Mr. McDowell."

"But what does it *mean*?"

"The math of quantum mechanics is crisp, beautiful, and wonderfully predictive. What is not clear," what not even Albert Einstein could discern, "is the physical meaning of that mathematical formalism. Some argue that to ask the question is impermissible. Some assert that the realm of quantum mechanics is so removed from our senses we're unequipped to judge." That, of course, was why Schrödinger devised the cat in the box. A cat is not a subatomic particle ....

Why did his mind keep wandering?

"There are several interpretations, all unprovable, of the mathematical formalism. Living or dead: To have but one outcome when the box is opened is unaesthetically asymmetric. Hence, one theory has it that both outcomes occur—which implies the spawning of another universe. More generally, whenever an uncertainty at the quantum level must resolve itself into a particular result, *the* universe itself must split into *many*, one to instantiate each possible outcome. If we, the occupants of one universe, unseal the box to let loose a live cat, in another universe, the occupants must encounter a dead feline."

More murmuring. This time Thaddeus let bewilderment run its course. As young minds grappled with countless myriads of branching universes newborn each moment, into Thaddeus's own churning mind popped the vision of two commingled universes. Of two possible professors in superposition. From what source might free will arise, except for quantum uncertainty?

*Children or not? Memories or hallucination? A bullet hole or just a hole?*

Clanging yanked his attention back to the lecture hall. An unseasonably warm autumn day and an alarm: almost surely a fire drill. "Attention, everyone! Leave in an orderly fashion. Assemble on the quad." Thaddeus watched the students stand, form lines, file efficiently from the room, his eyes sweeping from exit to exit to exit. His thoughts remained in turmoil.

"Mr. McDowell!" The lad was at the blackboard. Had he likewise concluded this must be a fire drill? "Cease your foolishness and go now." Thaddeus's eyes resumed their sweep. When his gaze next touched the front of the auditorium, the area was empty. A hastily scrawled phrase had appeared below Shrödinger's cat. He squinted to read it.

Now what had he been thinking about?

◆❖◆

The students filtered back from the sunny quad into the hall. A few glanced unsubtly at the wall clock. The hour was almost up. He could have dismissed them straight from the quad, instead of squeezing in a final few minutes of lecture.

"A pleasant day for a fire drill." Thaddeus picked up a piece of chalk. "Where were we?"

Tittering erupted as he looked to the flawless blackboard. His face, thankfully hidden from the class, reddened. "Quite clever." He briskly erased the scribbled graffito that had appeared beneath his crudely drawn sketch of Shrödinger's cat. The chuckling grew. "Very clever, indeed."

He wished he *had* dismissed them from the quad. A minute later the bell rang, ending the session. Grinning students in twos and threes bustled from the hall.

His humble drawing followed the student witticism into oblivion. Not that it mattered; the caption had been memorable enough. Straightening a sheaf of lecture notes, Thaddeus wondered whether even Einstein would have agreed.

"The *cat* knows."

---

The (mis)adventures of Thaddeus Fitch have caught others' eyes. Canadian film company Redhead Productions turned the story "Grandpa?" into the multiple-award-winning short film "The Grandfather Paradox."[1] "Grandpa?" was later redone as a podcast and, most recently, optioned by a second filmmaker.

Recently, *this* news came my way. "The Grandfather Paradox" has become part of the curriculum at the Los Angeles Film School in Hollywood, in a course that also includes the classic time-travel film *Back to the Future*. How cool is that?

As for "Inside the Box," I'll note that I would have preferred a different title. The editor of *Asimov's Science Fiction*, alas, doesn't care for puns.

Surely "Schrödinger's Pair O' Docs" would have fit the story just as well.

1. See http://www.redheadproductions.com/grandfather_page.html. This page lists the cast, crew, awards, and other basic info about the film. (But unless you click in from Canada— the Canadian National Screen Institute provided partial funding for the filming—the link to the movie itself won't stream anything.)

# *Out of the Cradle*

FTL travel to the stars isn't likely to happen anytime soon. What about baby steps, like going back to the moon, or visiting a near-Earth asteroid? How about Mars?

Growing up, I never doubted that humanity's future was in space. For a few years, the milestones came fast and furious: first artificial satellite (1957), first man in space (1961), first man on the moon (1969), and first orbital space station (1971).

End of milestones.

The space-shuttle fleet has been retired, with no NASA replacement in sight. Scarcely a month after the final shuttle flight, Russia suffered its fourth launch failure in nine months. The August 24, 2011 crash of a Russian supply ship brought into question the Russian Space Agency's capability to support the International Space Station.[1] The failed upper stage used to launch the Progress supply ship is also used to ferry crews to the ISS aboard Soyuz capsules. In October 2011, with the Russian rockets grounded, NASA was forced to make contingency plans to evacuate the ISS.[2] Fortunately, Russian engineers successfully diagnosed the problem. A replacement crew successfully docked with the ISS on November 14, 2011.[3]

1. http://news.yahoo.com/russian-supply-ship-space-station-crashes-172337040.html

2. http://www.spaceflightnow.com/station/exp28/110827unmannedops/

3. http://www.reuters.com/article/2011/11/16/russia-space-idUSL5E7MG1EX20111116

Are we seeing the end of the human-in-space era?

My friend and sometimes writing collaborator, Larry Niven, has commented that, "The dinosaurs became extinct because they didn't have a space program. And if we become extinct because we don't have a space program, it'll serve us right!" "Rock! Bye-bye, Baby," the essay that opens this section, considers how humans might be more proactive than our saurian predecessors.

Perhaps, just as in much early science fiction, commercial enterprises may lead the way. "Lost in Space? Follow the Money" looks at that prospect. It won't necessarily be easy: The same day as the recent Russian mishap, entrepreneurial rocket company Blue Origin lost a suborbital rocket.[4]

While this section offers only one piece of short fiction, I'll note that both essays capture research for an upcoming (as I write this) near-Earth, asteroid-related novel, *Energized.*

4. See "Amazon Chief's Spaceship Misfires," http://online.wsj.com/article/SB10001424053111904716604576546712416626614.html

# *Rock! Bye-Bye, Baby*

## (First published in November 2009)

In the early morning of December 14, 1807, something streaked across the sky above Weston, Connecticut and exploded. At least six rocky fragments reached Earth, the largest (now on display at the Yale Peabody Museum) weighing twenty-eight pounds. Yale University professors Benjamin Silliman and James L. Kingsley collected fragments and published a detailed report in the *Connecticut Herald.*

Upon first learning of the incident, President Jefferson said, "I would more easily believe that two Yankee professors would lie than that stones would fall from heaven."[1]

Times change.

That objects fall to Earth from the sky is now common knowledge.

This is (as I write) the hundredth anniversary year of the Tunguska Event, in which an aerial explosion (the site did not show an impact crater) from an asteroid or comet flattened and burned forest across a broad expanse of Siberia. Atmospheric disturbances extended to Western Europe, with Londoners able to read newspapers by the

1. This isn't a criticism of Jefferson, who was perhaps the most scientifically inclined of all American presidents. He not only chartered the Lewis and Clark expedition, he personally instructed Lewis on the information to be gathered and trained Lewis in surveying techniques. Long after (1962), while entertaining forty-nine Nobel Prize winners at the White House, President Kennedy said, "I think this is the most extraordinary collection of talent and of human knowledge that has ever been gathered together at the White House—with the possible exception of when Thomas Jefferson dined alone."

night sky's sudden eerie luminescence.[2] And an asteroid or comet strike is a leading explanation for the sudden disappearance of the dinosaurs—the so-called Cretaceous-Tertiary extinction event.

*Atlantic* magazine recently ran an article, "The Sky Is Falling."[3] That article was headlined with the provocative teaser: "The odds that a potentially devastating space rock will hit Earth this century may be as high as one in 10. So why isn't NASA trying harder to prevent catastrophe?"

Is Earth wearing a bull's eye? We'll look at that, relying on more primary sources, and at realistic approaches toward handling space-rock hazards. We'll also consider obstacles to preparing—and motivations beyond fear for developing—a planetary-defense capability.

First, let's get some definitions out of the way.

### *Meteoroids and asteroids and comets (oh, my)*

What *might* smack into Earth?

Many more things than planets (however defined—I won't go there) whiz about the solar system. We call these sub-planetary objects:

- meteoroids, if boulder-sized or smaller (and meteorites if they survive reentry and reach the Earth's surface), or
- asteroids, or,
- comets, if they exhibit a coma (gaseous envelope).

Meteoroids, by definition, are too small to endanger many people or any significant area. Only size distinguishes asteroids from meteoroids.

Asteroids mostly orbit between Mars and Jupiter, in the too inclusively named Asteroid Belt, at a distance from the sun of be-

2. "The Tunguska Mystery," Luca Gasperini, Enrico Bonatti, and Giuseppe Longo, *Scientific American*, June 2008. The event may have been more than an air burst. In the article, the authors assert that pieces of that exploding object survived to reach the ground, and that Lake Cheko was carved by such a fragment.

3. http://www.theatlantic.com/doc/200806/asteroids

tween about two and four astronomical units. (One astronomical unit, or AU, equals the mean distance between the Earth and the sun: about 93 million miles or 150 million kilometers.) Of course if *all* asteroids orbited in that belt, this would be a short article. And maybe dinosaurs would still rule the Earth.

The threat to Earth comes from Near Earth Objects (NEO). How near? The current standard is approaching within 1.3 AU of the sun. Stated another way, a NEO is anything that comes within 28 million miles of Earth's orbit.

Astronomers distinguish several populations of near-Earth asteroids. Generally speaking:

- The Aten asteroids orbit within Earth's orbit.
- The Amor asteroids approach from the outside, but do not cross, Earth's orbit.
- The Apollo asteroids have orbits that cross Earth's own orbit.

NEO can be comets rather than asteroids. Short-period comets originate in the Kuiper Belt, a disk-shaped region that begins beyond Neptune, about thirty AU from the sun, and extends (depending on whose guess you favor) to fifty or fifty-five AU from the sun. Long-period comets, it is hypothesized, originate in a spherical region even larger and more remote (and whose existence remains speculative): the Oort Cloud.

Asteroids and comets generally differ in composition. The ice line marks the distance from a star at which water can condense as ice. For our sun, the ice line is at about 2.7 AU—within the Asteroid Belt. Water evaporates from objects that orbit within the ice line, is driven outward by the solar wind, possibly to condense in the outer solar system. Other volatile materials (like ammonia and methane) behave in similar ways and have their own ice lines.

Objects originating in the inner solar system generally consist of rock or metal. Objects originating in the outer solar system add ices of various kinds. The latter objects are called comets. If the orbit of a comet is sufficiently eccentric—that's astronomer- and geometer-speak for deviating from circularity—to cross the ice line(s), sunlight starts turning the volatile materials to gas. Gases escaping the object's usually miniscule gravitational attraction, and any dust carried away by the gases, are pushed and shaped by the solar wind to form the comet's tail.

Most space-rock discussions focus on asteroids. Comets certainly can be dangerous, but their long, slow orbits are much harder to characterize. Known and suspected asteroids with Earth-approaching orbits far outnumber known and suspected comets with Earth-approaching orbits.

Not all objects represent the same threat. Size is a factor, obviously, but it's not the only factor. Speed of encounter (the net velocity difference between Earth and space rock) matters, naturally. Composition also matters. Metallic objects will better survive an atmospheric passage than stone. Solid objects survive better than loosely-packed rubble piles. And the angle of impact matters. Objects entering the atmosphere at a shallow angle undergo more friction from entry than rocks entering at a steep angle.

Solar-system objects do not have fixed orbits. True, many objects have *stable* orbits. They must, because the solar system has an age on the order of 4.5 billion years. The familiar planets have avoided a plunge into the sun or being hurled into the interstellar darkness. Still, forces are constantly at work on orbiting objects, including the gravitational perturbations from other orbiting bodies, tidal drag[4], the propulsive effects of gas eruptions, and light pressure.

When orbits change, things can collide ....

### *Expert opinions*

The Asteroid Deflection Research Symposium (ADRS), held October 23-24, 2008, in Arlington, VA, drew invited attendees from

4. Energy dissipated in ocean tides slows the Earth's rotation. To conserve the Earth-moon system's total angular momentum, the moon's orbital radius grows at about 4 cm/year. Similar effects slowly alter the orbits of other planet/moon systems.

concerned U.S. government agencies:

- the National Aeronautics and Space Administration,
- the National Research Council,
- the National Science Foundation,
- the Air Force,
- the Defense Threat Reduction Agency (part of the Department of Defense)
- the Department of Energy, and
- the Department of Homeland Security.

A small number of attendees came from academia (especially from the aerospace department of Iowa State University, organizer of the event), the aerospace industry, and science-oriented media outlets. A few were science-fiction authors like me.[5]

ADRS was hardly the first symposium on the topic of planetary defense, but it was *my* first. It appears to have been a good place to start. The organizers kept the conference small to facilitate candid discussion—and succeeded. Most attendees stressed that their presentations and participation reflected personal opinions and not the official viewpoints of their organizations. For that reason, I have not attributed information to individual participants.

ADRS provided the original impetus and much of the source material for this article.

### *How serious is the threat?*

The conference opened with an overview of the problem. Congress has tasked NASA to survey the threat from NEOs. NASA's Near Earth Objects Program maintains a running count of the findings at http://neo.jpl.nasa.gov/stats/. New NEOs are found more or less daily; discovery trends suggest the cumulative number of such objects

5. We SF types attended as members of SIGMA, a pro bono group of authors that offers the government our outside-the-box perspectives on a spectrum of issues. For more about SIGMA, see the group's website at http://www.sigmaforum.org/home.php.

will reach 6,000 in 2009. The count of large objects, a kilometer or more across, will likely reach 800 during 2009.

(Update: NASA's actual year-end 2009 tally was 6,642 NEOs, of which 805 are large. As of November 28, 2011, the count stood at 8,376, of which 832 are large.)

(This is as good a spot as any to stress that practically every number in this article is an estimate. I'll mention some major sources of uncertainty as I go, but a disclaimer before every number would get tedious. Different references, and the same reference at different times, may give different estimates. Legitimate scientific uncertainties do not detract from the overall message: lots of big rocks in the neighborhood that could do Earth—and us—harm.)

A kilometer—besides being a nice, round number—is about the size at which a rock's impact would have global repercussions. (One kilometer isn't yet dinosaur-killer sized: that impactor, whether an asteroid or a comet head—a question still in dispute—is estimated at about ten kilometers.) Initial effects would include fireball, air and ground shock waves, and (with a water impact) tsunami. Dust and debris would add acid rain, diminished sunlight, and secondary impacts.

How precise are the statistics? That's open to debate.

The search has used telescopes neither designed for asteroid hunting nor dedicated to the task.[6]

6. Current surveys involve narrow field-of-view astronomical instruments and mostly visible-light frequencies. The soon-to-come Panoramic Survey Telescope And Rapid Response System (Pan-STARRS) survey will take a different approach. The instrument nearing completion will have a 3° field of vision (wide for a large telescope), allowing full-sky surveys (of the sky visible from the Hawaii observatory) four times a month.

The instrument will also observe in the infrared band. Dark, faint objects that are difficult to spot by reflected visible-wavelength sunlight are expected to reveal themselves through the contrast between their sun-heated surface and the cold background of space.

(November 2011 update: The first of four Pan-STARRS telescopes entered operations in 2010. At full capacity, Pan-STARRS will operate a four-telescope array.)

(November 2011 update: NASA's Wide-field Infrared Survey Explorer

An object must be observed several times to derive its orbit. A "new" discovery may prove to be a repeat sighting of an object whose orbit had not yet been characterized or has shifted. The less massive an object, all others things being equal, the more likely it is that the object's orbit will change with time.

Objects this small aren't round, being too small for gravity to have collapsed them into a generally spherical shape. Glimpses of a rock from different sides differ, complicating the recognition, sizing, and orbital-determination problems.

Size estimates often derive from a pinpoint of reflected light. A dark rock is harder to spot than a light rock. Size estimates derive from estimates of both average asteroid albedo—we don't have up-close data about very many asteroids—and distance.[7] The sizing model will mistakenly conclude that a dark rock is smaller than it truly is.

Here's the good news: a decreasing rate of discovery for large NEO. This suggests that we've found most of them.

The not-so-good news: It doesn't take a kilometer-sized rock to do a lot of damage. Impact of a rock as "small" as 140 meters would cause a regional disaster. In 2005, Congress tasked NASA to find 90% of such medium-sized objects by 2020. Congress did not provide funding for this mandate; NASA's current forecast shows it missing the deadline by a few years.

The most worrisome NEO are dubbed Potentially Hazardous Objects (PHO) or Potentially Hazardous Asteroids (PHA)—two names for the same thing. By either label, these are large objects—

---

(WISE) orbital telescope was launched in late 2009 and decommissioned in early 2011. NASA reported of WISE, "Our instrument is finding [dozens] of asteroids every day. http://science.nasa.gov/science-news/science-at-nasa/2010/26mar_darkasteroids/)

7. *Albedo* is astronomers' term for the ratio of reflected light to incident light. An object whose albedo equals one is a perfect mirror. An object whose albedo equals zero is a perfect absorber. A black hole has an albedo of zero—by definition, light can't escape from a black hole.

Typical asteroid albedos are 0.05 to 0.2. See http://www.johnstonsarchive.net/astro/astalbedo.html

about 150 meters or greater[8]—whose orbits approach too close to Earth. "Too close" is defined as a 0.05 AU: about 4.65 million miles. In comparison, the mean Earth-moon distance is about 240 thousand miles.

NASA's NEO Program has identified about 1,000 PHO.

(As of late November 2011, the count is 1,270.[9])

Sky surveys probably won't give much warning of potentially dangerous comets (mostly Kuiper Belt Objects with highly eccentric orbits). Until a comet begins to show a significant coma or tail—i.e., as it gets closer to the sun—it is very difficult to spot and track. Even short-period comets have orbits of decades (the famous Halley's Comet has an orbital period of seventy-six years), so modern surveys have likely characterized very few of them.

The bigger the rock, the rarer, of course. Table 1, adapted from NASA's 2006 Near-Earth Object Survey and Deflection Study[10] and several ADRS presentations, gives one view (based on a variety of estimates, uncertainties, and assumptions) of the probabilities. The bigger the rock, all others things equal, the more dangerous. Estimates vary, but impact by an object as small as fifty meters might deliver five to nine megatons of energy.

---

8. That vagueness in PHO size threshold is of NASA's doing, not mine. The official PHO classification depends on the parameters from which sizes are inferred. "About" acknowledges the imprecision in size estimates.

9. http://neo.jpl.nasa.gov/neo/groups.html

10. http://www.spaceref.com/news/viewpr.html?pid=22583

| Type of Event | Diameter of Object | Fatalities per Impact (Note A) | Typical Rate (Years Between Events) | Fatalities (In 100 years) (Note B) |
|---|---|---|---|---|
| High altitude break-up | < 50 m | ~0 | Annual | 0 |
| Tunguska-like event | > 50 m | ~5000 | 250 - 500 | 3,400 |
| Regional event | > 140 m | ~50,000 | 5,000 | 2,700 |
| Large sub-global event | > 300 m | ~500,000 | 25,000 | 5,400 |
| Low global effect | > 600 m | > 5 million | 70,000 | 19,400 |
| Nominal global effect | >1 km | > 1 billion | 1 million | 282,000 |
| High global effect | > 5 km | > 2 billion | 6 million | 90,000 |
| Extinction-class event | > 10 km | 6 billion | 100 million | 16,000 |

**Table 1.** Risks of, and from, asteroid/comet encounters

Table notes:

(A)In a topic rife with assumptions, the metric *fatalities per impact* is among the most subjective. The Tunguska event of 1908 had no known human casualties—because it happened in a remote part of Siberia. The area devastated by the event is larger than metropolitan Washington or New York. Location matters.

(B)The "Fatalities (in 100 Years)" entries reflect a Poisson distribution of collision probabilities around the mean estimated arrival rate given in the "Typical Rate (Years Between Events)" column.

And that alarming forecast in *Atlantic?* Sky surveys do not back it up. Nor does the historical record. Nor, in fact, does the article itself. Its only substantiation is an expert's statement citing—without context, definition, or explanation—"a one-in-10 chance per century of a dangerous space-object strike." For *some* level of danger, that is certainly true.

### *A case in point*

When first discovered in 2004, the asteroid now named Apophis 99942 was given a 2.7% probability of hitting Earth in 2029. In its longest dimension, Apophis is believed to measure about 220 meters—which, as Table 1 suggests, would make its impact unpleasant. Refined orbital estimates for 2029 now predict Apophis won't come any closer than 18,300 miles—from the Earth's *center*. That is, Apophis might pass inside the orbits of the geosynchronous communication satellites. (Because of the angle of approach, the satellites aren't at risk.)

As for Apophis' return visit in 2036, there remains an estimated 1-in-45,000 chance of a collision with Earth.

Why state collision risks as probabilities? First, orbit determinations are statistically derived from NEO sightings whose estimates of position and velocity unavoidably include a measure of uncertainty. Second, orbits can and do change over time, influenced by rarely known details like mass, spin, absorption and reemission of sunlight, and the gravitational influence of other asteroids. Even the influences of major planets are inexact since planetary orbits themselves are known only through estimates. The uncertainty in Apophis' nearest approach in 2036 is millions of miles. The most likely separation that year is a comfortable 0.32 AU.[11]

### *A small success*

NASA's rock-tracking prowess recently had its fifteen minutes of fame. On October 6, 2008, the NASA-funded Catalina Sky Survey

---

11. http://neo.jpl.nasa.gov/apophis/

identified a small (few meters across) NEO dubbed 2008 TC 3 and predicted its atmospheric entry over Sudan the next day. The object did, in fact, burn up in the atmosphere over Sudan exactly as predicted, releasing an estimated one kiloton of energy.[12]

### *When push comes to shove*

If an asteroid or comet is headed our way, there is really only one thing to do: Shove it into a slightly different orbit. The difficulty, of course, is that an object large enough to be dangerous carries a *lot* of momentum.

Let's consider the mass first. As one idealized asteroid, we'll assume a sphere fifty meters in diameter, with the density of a typical stony meteoroid: about 3.5 grams/cc. (Spheres make for easy calculations. The term often used to describe actual asteroids is "potato-shaped.") This sphere masses about 229 million kilograms, or about 500 million pounds. As a second example, consider a 150-meter-diameter sphere—barely meeting the size standard for a PHO. This time we'll assume a composition like an iron meteoroid: 8 gm/cc. *This* sphere masses about 14.1 billion kilograms, or a bit over 31 billion pounds.

Next, we'll consider velocity. Meteoroids typically reach Earth with velocities in the range of 10 - 70 kilometers/second (or about 6 - 43 miles/second). Earth's atmosphere slows most meteoroids down to terminal velocity, but would hardly affect anything as massive as our example asteroids.

Consider a mass *M* traveling at velocity *V*. (For simplicity we'll assume the object isn't spinning. Spin would contribute angular momentum and rotational kinetic energy.) Elementary mechanics gives us:

$$\text{Momentum} = M*V$$
$$\text{Kinetic energy} = 0.5*M*V^2$$

The squared term in the kinetic-energy formula means the total energy goes up rapidly as velocity increases. Twice the velocity

12. http://www.planetary.org/blog/article/00001684/.

means four times the kinetic energy. When the asteroid comes to a halt—stopped in its tracks by the far more massive Earth—that kinetic energy turns to heat and shock waves.

Table 2 shows the momentum and kinetic energy involved if our sample asteroids strike with representative velocities. For comparison, there's a row for a fully loaded tractor trailer traveling 65 miles/hour. The values for momentum and kinetic energy are shown in scientific notation—millions and billions hardly suffice. The kinetic energy of impact is also shown in equivalent megatons of explosives.

For reference, the largest atomic weapon ever tested had a yield of fifty megatons.

| Object | Mass | Velocity | Momentum | Kinetic energy |
|---|---|---|---|---|
| Tractor-trailer (max. legal weight) | 36 thousand kg | 0.03 km/sec (65 mi/hour) | $1.1 \times 10^{6}$ kg m/sec | $1.53 \times 10^{7}$ joules (~3.65 billionths megaton) |
| 50-meter stony asteroid | 229 million kg | 10 km/sec | $2.3 \times 10^{12}$ kg m/sec | $1.15 \times 10^{16}$ joules (~2.74 megatons) |
| 150-meter iron asteroid | 14.1 billion kg | 70 km/sec | $9.9 \times 10^{14}$ kg m/sec | $3.46 \times 10^{19}$ joules (~8,278 megatons) |

**Table 2.** Representative collisions

### *When the sky really* is *falling*

To deflect an imminently dangerous (per Table 2, read: massive and fast approaching) asteroid requires a large push. Conclusions at ADRS differed on many topics, but on one there was near unanimity: Nuclear explosives are the technically superior—and in many cases, the only currently practical—defense against an imminent PHO strike. Intuitively, that's reasonable. Shifting the orbit of megatons of rock or metal will take megatons of energy.

In principle, nuclear deflection is straightforward. A nuclear detonation near (or within) an asteroid releases a flood of gamma rays (very energetic photons) and neutrons. The radiation vaporizes part of the asteroid; the spewing gases create thrust. The radiation itself also delivers a push.

Alas, we know so *little* about asteroids, and that ignorance creates uncertainties about the effects of an explosion. The radiation absorption depends on the elemental composition of the asteroid. The asteroid's composition also influences what fraction of neutrons is blocked—absorbed by material already boiled off by the faster-moving photons. A blast might break an asteroid into pieces still headed our way, many still large enough to be dangerous. Or the fragments might coalesce over time, drawn together by their mutual gravitational attraction.

Delivering a nuclear device adds complication. The asteroid is a fast-moving target. Putting a spacecraft at precisely the correct spot and detonating the device at exactly the right instant as the spacecraft speeds past stretches the limits of navigational practicality.

The ideal detonation position is far from obvious. While ADRS presentations considered optimum detonation positions for idealized asteroids (and results varied), real asteroids won't be ideal. Optimization should take into account the target's shape and composition. An ill-timed detonation near an asymmetric object will waste energy by uselessly altering the object's spin.

Split-second timing for a flyby detonation is not the only approach. Another option is a rendezvous mission, whether to take up a position near the object, make a landing, or (with a multipart payload) both. On the plus side, a rendezvous allows close-up study before making the targeting decision (or decisions, because the spacecraft might bring multiple devices). On the negative side, decelerating to orbit or land on the PHO expends more fuel and time. Carrying more fuel means bringing a smaller payload—hence, all other things equal, less ability to deflect the PHO. If the target is a comet, the coma is both a barrier to observation and a source of debris potentially hazardous to the spacecraft. Loitering near a comet might reduce the odds for mission success.

If a PHO on a threatening trajectory allows us more lead time (i.e., if we discover the PHO and characterize its orbit early enough), the mission constraints are eased. A small orbital change made well ahead of time may suffice. In this case, the payload might simply be a high-energy impact device. Like bullet-hitting-a-bullet missile defense—and ABM programs deal with a closing rate of "only" about 8 km/sec—direct impact is a tough problem. A near-miss with a nuke has some effect; a near-miss by a kinetic impactor does nothing. Like a nuke, any major explosion or impact risks fragmenting the object.

### *Nudge, nudge*

What if a PHO allows us lots of lead time? Then more subtle methods may suffice to change the object's orbit. Even sunlight, given enough time, might do the trick—just paint the object white.[13]

Slightly faster-acting would be a structure that alters the sunlight pressure on the NEO. A huge solar sail, perhaps configured as a parabolic reflector (to concentrate the sunlight), is one approach. A solar sail kept between the object and the sun becomes a solar shade; it reduces the incident light. Such structures must be huge to have much effect, even over long time spans. Huge here means hundreds of meters in diameter. The structure will be tricky to deploy and, however thin the material, massive. And while the science of solar

13. The Yarkovsky effect describes the force on a rotating body from differential emission of thermal photons. The sunlit side of an asteroid is hotter than the dark side, making the emission of heat—thermal photons—asymmetric. See http://en.wikipedia.org/wiki/Yarkovsky_effect.

sails is well-established, solar-sail propulsion remains—after several attempts—an undemonstrated technology.

One could install a propulsion device on an asteroid. It's unlikely we could deliver enough fuel to appreciably change an asteroid's orbit, but perhaps fuel could be produced in situ. Or perhaps solar cells can power an electromagnetic rail gun, using parts of the asteroid itself as reaction mass. Or a solar-powered laser on a standoff spacecraft could boil off asteroid material.

The gravity tractor is a particularly elegant concept. Place a small payload, the tractor, near the PHO. Gravity draws tractor and asteroid toward each other. Now, slooooowly—so as not to overcome the weak attraction—move the tractor. The asteroid follows. The gravity tractor operates over a long period of time, so low-thrust, solar-powered propulsion is needed.

The tractor's propulsion could come from a solar sail or ion thrusters powered by an array of solar cells. (Ion thrusters use electrical fields to expel a high-speed stream of charged particles, typically xenon ions. Ion thrusters produce low acceleration but can maintain thrust for much longer than chemical rockets.)

Using solar-sail propulsion, there is only one position for the gravity tractor: where light-pressure thrust on the sail balances the gravitational drag between asteroid and tractor. That positional restriction makes this sort of gravity tractor a single point of failure. If anything should go wrong, we lose our defense against the object that was being towed. A single ion-thruster-propelled gravity tractor is also a single point of failure. A constellation of solar-cell-equipped, ion-propelled gravity tractors, however, would have redundancy if single units failed.

So a dramatic intervention or (time permitting) something subtle? On the one hand, noncontact methods work very slowly. The mechanism must work flawlessly—or be deployed with sufficient redundancy to cope with failures—for many years. On the other hand, noncontact methods avoid the risk and uncertainties of fragmenting the PHO. Given how little we know about specific asteroids and comets, that is a major plus.

With so much relevant technology yet to be proven, and so few asteroids and comets yet explored, it's premature to commit to one or two defensive measures. We cannot yet know in detail how space objects would respond to nuclear detonations or kinetic impacts, or whether in situ resources can be used to change space objects' orbits gradually.

***Progress***

Enough dwelling on what we don't know. We've learned a great deal about asteroids and comets in the past few years ago, thanks to:

- NEAR-Shoemaker: rendezvous and landing mission to asteroid Eros.
- Deep Impact: rendezvous and collision with Comet Tempel 1.
- Stardust: sample return mission to Comet Wild 2.

And we'll learn even more about asteroids and comets in the next few years:

- Dawn: mission to asteroid Vesta and (now promoted to dwarf planet) Ceres, arriving at the former in 2011 and the latter in 2015. (Update: Dawn achieved orbit around Vesta on July 16, 2011.)
- Rosetta: orbiter/lander mission to Comet 67P/Churyumov-Gerasimenko, arriving in 2014, after two asteroid flybys (European Space Agency).
- (Update: In May 2011, NASA announced the OSIRIS-Rex probe: Launching in 2016, the mission aims to return an asteroid sample to Earth in 2023.[14])

14. http://www.washingtonpost.com/national/nasa-mission-will-kiss-an-asteroid-to-get-vital-info-about-lifes-origins/2011/05/25/AGnTPXBH_story.html

### *Getting there is half the fun*

All the mitigation options make a common assumption: that we can deliver a significant payload to the PHO. For an object closing on Earth at 70 km/sec, that's no easy task. Almost certainly, chemical rockets alone won't suffice for missions to such fast-moving rocks. For reference, the Saturn V rockets used in the Apollo program cut off at 11.2 km/sec.

Over time, spacecraft with ion thrusters can build up enormous velocities. In principle, so can solar sails. The catch is, time permitting.

Gravitational slingshots are another way to gain velocity. The Rosetta mission, for example, will have swung around Earth three times before it reaches its destination comet. Exploiting Earth's (or another body's) gravitational energy in this way costs maneuvering time, and requires that some planet be in the right place at the right time.

### *The real world intrudes*

It's worth remembering: For lead times of years (or perhaps even a few decades), standoff nuclear devices may offer our only chance of deflecting kilometer-scale objects.

The Outer Space Treaty[15], in force since 1967, prohibits the stationing of nuclear weapons in space or on any celestial object. The treaty has no loophole to permit use against a PHO. (One ADRS attendee argued unpersuasively that deflecting a space object was "propulsion," not a weapon, and thus does not come within the purview of the treaty.)

Even more restrictive for planetary defense, the treaty expressly forbids "the testing of any type of weapon" in space. I'm no space lawyer, but kinetic impactors—guided projectiles with closing speeds of kilometers/second—seem like weapons to me. If they are to become a part of our planetary defenses, I would *hope* we test them before needing one to work.

15. Formally known as The Treaty on Principles Governing the Activities of States in the Exploration and Use of Outer Space, Including the Moon and Other Celestial Bodies. See http://www.state.gov/t/isn/5181.htm

Could the treaty be amended? Clarified? Surely countries would rather update (or ignore) the treaty than face a major asteroid or comet strike. Whether the Outer Space Treaty should be amended in anticipation—say, to allow a demonstration test against a harmless asteroid—seems more contentious.

As with most international treaties, signatories can withdraw upon one year's notice.

Judging from ADRS discussion of planetary-defense policy—and here we debated as citizens, not technical experts—only the questions are clear. Would countries agree to subcontract Earth's defense to a particular space-capable nation(s)? What is to stop a country from acting unilaterally? While an international planetary-defense organization might be preferred by some nations, other countries would be loath to see an international organization controlling missiles and nukes. Will the public trust an international entity to maintain control over its nukes, or not to preempt its nukes for other purposes? If a NEO intervention is only partially successful—the asteroid, or a fragment thereof, smacks Country B instead of Country A—who is responsible? What if a defensive launch fails and spreads radioactive materials? Groups have taken NASA to court (so far, without success) to stop launches of spacecraft using radioisotope power generation. Might someone sue an international planetary-defense group, perhaps in the World Court, to prohibit launches of nuclear devices?

Some simulations suggest that several small nuclear detonations would deflect an object better than one big blast. The organization(s) entrusted with Earth's defense might therefore want to develop or stockpile miniature nuclear devices: "suitcase nukes." The associated proliferation risk is not insignificant, and one terrorist attack using a stolen suitcase nuke would produce a casualty count comparable to a fairly hefty asteroid.

Maybe we won't seriously prepare for an asteroid strike. The basic data are in Table 1: While fatalities could be cataclysmic, the expected fatalities in a century—fatalities weighted by the likelihood of an impact—are quite small. Other challenges, like diseases and climate change and the current recession, have more severe expected short-term consequences and are apt to consume mindshare and scarce threat-mitigation resources. Like not buying insurance against a "once-in-a-century" flood risk, skimping on planetary defense may look prudent—until it doesn't.

If we *knew* a big rock would hit Earth, our attention would likely increase. We could chose to wait until we spot such a rock and hope the world will then respond wisely—but to wait guarantees we'll have to respond to the emergency from a state of ignorance and unpreparedness.

### *What now?*

Let's think proactively. What preparations might we take for planetary defense? At a high level, there are three tasks to undertake:

- Study—and tag with a radio beacon, for more accurate tracking—a few asteroids chosen for their threat potential. (If they are of scientific interest, so much the better.) Without detailed knowledge of the mass, shape, and composition of an object, any deflection plan will necessarily be subject to major errors.
- Test—on non-threatening asteroids—candidate deflection technologies.
- Decide who has what planetary-defense authority, and amend the Outer Space Treaty to allow for planetary defense.

We almost certainly have decades before a NEO endangers us, but most mitigation strategies will also need a long time to deploy and operate. It's risky to ignore the NEO hazard until a collision is near certain.

If you consider planetary defense worthy of investment, what can you do?

Option 1: Lobby NASA and Congress. It may help to point out that PHO-related missions will also advance the main areas of NASA focus: technology demonstration and science. Nor need you lobby alone. Consider joining a space-advocacy group, such as the Planetary Society. Or you might join an asteroid-threat-specific organization, such as—named for the asteroid in Antoine de Saint-Exupéry's charming *The Little Prince*—the B612 Foundation. (Such groups do more than lobby. The two I named also support the development of relevant technologies.)

Option 2: Convince a billionaire. Wealthy citizens are massively underwriting such worthy research projects as the Large Synoptic Survey Telescope Project (which has many uses, including asteroid surveys) and the Allen Telescope Array (for SETI purposes).

For years, NASA has had a fairly stable budget (well below that of its Apollo heyday), and existing programs have their constituencies. Synergizing with current NASA initiatives may be a more realistic goal than competing for funds. Crewed spaceflight programs are among NASA's biggest efforts, and can contribute.

A near-Earth-asteroid mission is in some ways—notably the fuel payload needed for target rendezvous and return to Earth—easier than a moon mission. Announcing a NEO destination for the Constellation program (NASA's latest human-spaceflight initiative) rather than a lunar return mission might spare the U.S. from an embarrassing space-race loss to China.

(November 2011 update: NASA's Constellation program has been canceled; an asteroid mission has become a likely early destination for the successor manned-spaceflight program. I'll have more to say about this in the next likewise updated essay, "Lost in Space? Follow the Money.")

A mission to Phobos or Deimos—both are likely asteroids captured by Mars—would be easier than a mission to Mars itself. From a base on either Martian moon astronauts could prove most of the technology necessary for a Mars base. They could also teleoperate robot explorers[16] without contaminating the pre-biotic conditions—or perhaps, even, alien life—on Mars itself.

16. Phobos orbits within 10 thousand kilometers of the center of Mars. Light-speed communication delay between Phobos and a robot on the Martian surface would be negligible.

The current Mars rovers are terrific, but there is a reason—Earth/Mars comm delays—why their explorations are measured in months per mile. With real-time control, robots could explore much farther and faster.

And trying to be positive, in 2008 Congress tasked NASA to recommend "the optimal approach to developing a deflection capability." Now if Congress would only increase the $4 million NASA spends annually on asteroid surveys and deflection studies ....[17]

### *Catch a falling star*

Asteroids are potential resources, too.

A few near-Earth objects of suitable composition, nudged into safe and convenient orbits, could revolutionize the economics of spaceflight and space colonization. Water, of course, from comets. Oxygen from the water. Metals refined in space using plentiful solar power. Silicon processed in space to build mass quantities of solar cells.

Cheaper spaceflight and in-space resources could also revolutionize our terrestrial energy infrastructure. Earth's atmosphere blocks most sunlight but is almost transparent to microwaves. Solar-power satellites beaming microwaves to antenna farms on the ground will be much more efficient than surface-based solar-power generation.

To study, reach, and change the orbit of a NEO will take a significant investment. Perhaps the fortunes to be made will make a more compelling case for that investment than fear of an unlikely—but devastating if it happens—collision. Of course, NEO exploitation, like NEO deflection, raises issues of trust. Whom do you trust to aim an asteroid *toward* Earth?

Fear and greed are great motivators. Here's hoping at least one of them motivates humanity enough to avoid the dinosaurs' fate.

17. The 1998 disaster movie *Armageddon*, about an asteroid threatening Earth, had a reported budget of $140 million. *Deep Impact*, same year, same subject, had a reported budget of $75 million. Our priorities seem wrong.

***To read further***

NASA FAQ page on asteroid and comet impact hazards, http://impact.arc.nasa.gov/intro_faq.cfm

NASA's Near Earth Object Program, http://neo.jpl.nasa.gov/

Minor Planet Center (the clearinghouse of record for asteroids and comets, including their designations and orbits), http://www.cfa.harvard.edu/iau/mpc.html

B612 Foundation, http://www.b612foundation.org/. Its mission is "to find potentially threatening asteroids, track their trajectories, and to demonstrate the technology to alter the orbit of an asteroid in a controlled manner." B612 counts several astronauts among its members.

1st IAA Planetary Defense Conference: Protecting Earth from Asteroids, http://www.aero.org/conferences/planetarydefense/index.html

Lifeboat Foundation (dedicated to "encouraging scientific advancements while helping humanity survive existential risks") Asteroid Shield white paper, http://lifeboat.com/ex/asteroid.shield?white

# *Lost in Space? Follow the Money*

## (First published in October 2011)

Explorer and Vanguard. Mercury, Gemini, and Apollo. Space shuttle and International Space Station. Every one an American space-age accomplishment—and, moreover, a government accomplishment.

(Yes, I know: it's the *International* Space Station. That said, NASA conceived, led, and largely paid for the project. As much as other nations contributed, the essential partner was the U.S.)

The government part is about to change.

As I write this, in October 2010, Congress has just passed a bill reauthorizing NASA for three more fiscal years. Per that reauthorization, NASA's shuttle fleet will retire in 2011, after only three more flights. Thereafter, for several years, NASA will pay the Russian space agency to deliver American astronauts to the International Space Station (ISS). NASA will pay Russia and other spacefaring countries to deliver cargo to the ISS.

Constellation, NASA's post-shuttle crewed spaceflight program—conceived of to return humans to the moon and, eventually, to send humans to Mars—is in a wind-down stage.[1] NASA's long-

1. The Obama administration originally called for Constellation's complete cancellation; Congress and the administration continue to negotiate the details of program termination. Some elements (e.g., the Orion crew capsule, repurposed as an ISS emergency escape capsule), seem likely to survive. Stay tuned for the annual budget appropriation process.

November 2011 update: As above, the Constellation program was canceled, but Constellation's Orion crew capsule lives on as the Orion Multi-Purpose Crew Vehicle. NASA envisions an unmanned test flight of Orion MPCV in 2014. (For general program information, see http://www.nasa.gov/exploration/systems/mpcv/index.html.)

term thinking will turn to possible explorations *beyond* the moon. To that end, in 2015 NASA is to begin development of a heavy-lift vehicle, with an eye toward a manned asteroid rendezvous mission by 2025 and a manned Mars mission in the 2030s.[2]

Grand ambitions—at a leisurely pace, without near-term goals—while, for several years, at least, NASA will lack any crewed spaceflight capability.

Has the United States charted a bold new course? Or have we lost our way in space?

NASA's latest roadmap begins a transition to a new reliance on commercial services for near-Earth space operations. NASA's newly approved budget allots $1.3 billion over three years to encourage development of commercial spacecraft (versus $3.3 billion in the Obama administration's original budget request). The space agency's new approach won't succeed unless its new, commercial partners do. And so, we need to understand the status and prospects of the commercial-space business.

(A few article-wide disclaimers: When a space-pioneering company you expected to find herein goes unnamed, it's likely because, for brevity, I've had to limit myself to examples. Similarly, companies may only be mentioned for some of their activities. Finally, the aerospace sector is very dynamic; the hot new company I write about today may, by the time this article sees print, have gone the

---

2. For more about NASA's long-range plans as embodied in its latest authorization bill, see "NASA in Transition as Congress OKs New Direction," Clara Moskowitz, September 30, 2010, Space.com (http://www.space.com/news/nasa-transition-congress-new-direction-100930.html) and "Mars, Here We Come! Congress Approves $19 Billion NASA Budget," Tariq Malik, September 30, 2010, Space.com (syndicated at http://www.foxnews.com/scitech/2010/09/30/mars-asteroid-congress-approves-nasa-budget/).

November 2011 update: The launch-system specifics of the post-Shuttle, post-Constellation era have become clearer. NASA has begun development of a new, crew-rated family of launchers called the Space Launch System (SLS). The largest SLS launcher will have a payload capacity of 130 metric tons—more than a Space Shuttle or (by a hair) a Saturn V could manage.

See http://www.nasa.gov/exploration/new_space_enterprise/sls_mpcv/index.html.

way of, say, Eastern Air Lines, defunct, or Douglas Aircraft, folded into a successor company.)

### *The redirection of NASA: good idea or bad?*

Surprise: Opinions differ.

Charles F. Bolden, Jr., NASA's current administrator, naturally supports his organization's current plan. His immediate predecessor, Michael Griffin, has staked out—to put it mildly—an opposing position. " 'One day it will be like commercial airline travel, just not yet,' Griffin said. 'It's like 1920. Lindbergh hasn't flown the Atlantic, and they're trying to sell 747s to Pan Am.' "[3]

Nor does the astronaut corps agree whether NASA's new concept is for good or ill. Twenty-four astronauts sent a letter to Congress backing NASA's new direction.[4] Other astronauts hold the opposite opinion, including Apollo veterans Neil Armstrong, James Lovell, and Eugene Cernan. Their open letter to President Obama warns that under current plans, "The USA is far too likely to be on a long downhill slide to mediocrity."[5]

At issue are very basic questions. First, will the American space program suffer a devastating loss of purpose and expertise while, for several years, NASA is an agency with neither an operational human spaceflight capability nor one under active development?

Second, after the Constellation program ends in favor of some ill-defined better future technology, will an unfortunate precedent have been set? That is, might NASA's *next* proposed flagship program

---

3. "NASA budget for 2011 eliminates funds for manned lunar missions," Joel Achenbach, February 1, 2010, *Washington Post*,

   http://www.washingtonpost.com/wp-dyn/content/article/2010/01/31/AR2010013101058.html

4. http://www.dailygalaxy.com/my_weblog/2010/09/-nasa-should-use-private-spaceships-say-astronauts.html

5. http://www.msnbc.msn.com/id/36470363

(whatever it ends up being) in turn be made to defer to enticing possibilities even further out? What about the program after that? And the one after that? (As a former boss once advised me, "The best is the enemy of the good enough.")

These questions are, at their core, political and sociological, and I won't attempt to answer them. This article focuses on the root question about NASA's new direction: *can* industry step up to an expanded role in space? Because if not, the new approach will certainly fail.

Short of waiting years to see how things wind up, how do we decide if a more heavily commercialized space program is a good or a bad idea? In futurism as in detective fiction, a good rule is: follow the money.

### *A brief history of commercialized space*

(Emphasis on *brief*. This article surveys a broad and dynamic topic—metaphorically the view from low Earth orbit. When the view leaves you wanting to see more, try Google. That will only be fair: Google uses data from LEO, too.)

At one level, American spaceflight has always been commercial. From Day One, NASA obtained the services of such aerospace contractors as:

- McDonnell (Mercury and Gemini capsules)
- Grumman (the Apollo Lunar Module)
- North American Aviation (the Apollo Command/Service Module)
- Boeing, North American Aviation, and Douglas (the first, second, and third stages, respectively, of the Saturn V launcher)
- Hughes Aircraft Company (the Galileo Jupiter probe)

Although each of these projects succeeded technically, only Boeing—on the strength of its aircraft business—has survived as an independent business. To pursue the largest space opportunities, even giant companies have seen the value of partnerships. We thus have the United Launch Alliance, a joint venture of Boeing and Lockheed Martin, to provide launch services to the U.S. government. The United Space Alliance—same partners, different mission, better acronym—operates the space-shuttle fleet.

Bidding for government space contracts is not without its risks. As we will see, chasing government contracts is not the only space business model.

In the more than half century since Sputnik, something quite remarkable has happened to commercialize space business: privately owned satellites, especially communications satellites. Indeed, the first privately sponsored space launch, way back in 1962, was to orbit a comsat: Telstar, funded by AT&T (with government partners, of course).[6]

Modern commercial comsats support telephony, radio, and television. Service providers reliant on comsats include Iridium Communications LLC (satellite-based, global mobile phone service), Sirius XM Radio (satellite-based radio distribution to cars and trucks), DirecTV (direct-to-home video distribution), and Intelsat (fixed-point communications services, such as distribution to local broadcasters and cable companies of network-television feeds).[7]

After decades of experience, hundreds of comsats, and millions of customers, providing space-based communications services remains a tricky business. Iridium, the comsat-based global telephone company, went from a darling of Wall Street to a multi-billion-dollar blunder to a niche company resurrected out of bankruptcy.[8] Merely launching comsats can be problematical, too. Take Sea Launch, the

6. http://en.wikipedia.org/wiki/Telstar

7. Intelsat began life as an intergovernmental consortium. It became a privately held company in 2007.

8. When Iridium filed for bankruptcy in 1999, it was among the twenty largest bankruptcy filings in U.S. history. See "Learning from Corporate Mistakes: The Rise and Fall of Iridium," Sydney Finkelstein and Shade H. Sanford. See the full article at http://mba.tuck.dartmouth.edu/pages/faculty/syd.finkelstein/articles/Iridium.pdf.

international consortium led by Boeing and Energia (the Russian rocket company) to launch satellites primarily for the comsat market. As I write, Sea Launch is in Chapter 11 bankruptcy.[9]

There is more to commercialized space than comm. Global positioning (and related applications, like truck-fleet management) is a huge space-based application. Glossing over the mathematical complexities, a GPS receiver calculates its three-dimensional location by comparing the arrival times of very precise clock readouts broadcast by a set of satellites. Knowing an exact position at all times is handy—especially for soldiers in the desert or when piloting unmanned aerial vehicles. Not surprisingly, the original GPS satellite constellation is owned and operated by the U.S. Department of Defense.

(Again not surprisingly, the GPS satellite broadcasts that make possible unclassified GPS applications are degraded to reduce their accuracy. The military can even suppress the civilian-accessible signals. Other countries want positioning systems on which they, too, can rely in a crisis. The Russian equivalent to GPS is called GLONASS. The European Union and China have global systems under development, and several countries are investing in regional satellite-based positioning systems.)

Had tax dollars not paid for the design, construction, and launching of global positioning satellites, that handy geolocation device in your car—and the geo-aware cell phone in your pocket—would be much less affordable.

Then there's remote sensing. The NASA activity initially known as the Earth Resources Observation Satellites Program launched the first Landsat (we're up to number seven) in 1972. Remotely sensed Earth data has commercial applications ranging from cartography to weather forecasting, agriculture to futures trading, mineral exploration to surveillance. Landsat 7, launched in 1999, was the last of the program—but not the latest source of remote-sensed data. Data can

9. http://www.spacenews.com/launch/100312-sea-launch-preparing-bankruptcy-exit.html

now be bought from commercial remote-sensing-satellite operators like GeoEye, DigitalGlobe, and the French company Spot Image.

By one tabulation[10], global space business exceeded $260 billion in 2009—growing 7% in a tough, recessionary year. In approximate thirds, that spending sorts into:

- $90.6 billion in commercial satellite services,
- $86.2 billion in global government space spending ($64.4 billion of that spent by the U. S. government), and
- $83.6 billion in infrastructure (e.g., launch services, spacecraft manufacture, and ground equipment).

### *To orbit or bust*

Given that corporate giants like Boeing and Lockheed Martin found it expedient to partner to operate the NASA space shuttles, can NASA realistically expect industry to step up to developing a new generation of launchers and capsules to deliver cargo or crew to orbit?

Apparently, because new players were already rushing into the field.

Take Space Exploration Technologies Corporation (or SpaceX, as it is more commonly known), started by PayPal cofounder Elon Musk. In 2006, SpaceX won one of the competed slots in NASA's Commercial Orbital Transportation Services (COTS) program.[11] COTS makes milestone payments for progress toward private spacecraft able to deliver cargo to the ISS. In 2008, SpaceX won an award under NASA's related Commercial Resupply Services (CRS) program[12] for twelve deliveries, each of at least twenty metric tons, to the ISS. In 2010, SpaceX successfully orbited a dummy payload

---

10. http://www.reuters.com/article/idUSTRE63B40C20100412

11. http://www.nasa.gov/exploration/news/COTS_selection.html

12. http://www.nasa.gov/home/hqnews/2008/dec/HQ_C08-069_ISS_Resupply.html

on the first test of its Falcon 9 launcher. Pretty impressive for a company established in 2002.

The COTS program is not for the faint of heart. Program participant Rocketplane-Kistler went out of business in 2010, and NASA transferred the R-K contract to Orbital Sciences Corporation. In 2008, Orbital was also awarded a CRS contract (in this case, for eight flights to the ISS). Together, the SpaceX and Orbital resupply contracts are valued at $3.5 billion, with potential to grow to $6 billion.

NASA's third commercialization activity, still at the conceptual stage, is the Commercial Crew Development program.[13] CCdev envisions privately operated human-rated flights to the ISS.

There is precedent for the U.S. government financially backing development of an innovative private transportation system. A century ago the Postal Service, contracting with private carriers to establish airmail service, was a major early backer of commercial aviation.[14] Here's hoping NASA's new policy is as successful.[15]

### *Coffee, tea, or space-sick bag?*

The big near-term commercial opportunity in delivering people to space won't be the occasional NASA-chartered trip to the ISS—it'll be passengers on private business. Rather than "space tourist," some prefer the term "citizen astronauts." Fair enough: as we'll see, reasons abound for starting a space line. Transporting sightseers is merely one.

That said, space tourism is the person-in-space business model closest to getting a serious test.

The age of space tourism has already begun—for the (extremely)

13. http://www.spacenews.com/civil/101001-nasa-bids-new-round-ccdev.html

14. http://inventors.about.com/library/inventors/blmailus3.htm#AIR

15. Whereas we wouldn't want to emulate the experiment called Missile Mail. See http://inventors.about.com/library/inventors/blmailus4a.htm.

well-to-do. Space Adventures, Ltd., contracting with the Russian space agency, has sent seven private astronauts aboard Russian Soyuz capsules to the ISS. Dennis Tito, the first private astronaut, reportedly paid $20 million for his trip.[16]

Are you a few million short? For less than $5,000, you can still experience weightlessness—if only for less than a minute at a time—flying maneuvers on a Zero G Corporation aircraft. (Space Adventures acquired Zero G Corporation in 2008. Parent and subsidiary are both US-based.) Zero G clients have included Stephen Hawking and Martha Stewart.

Today, there's nothing between. But that's about to change.

### *Space tourism—the shape of things to come*

Perhaps the most visible next-generation space-tourism company is Virgin Galactic. Backed by flamboyant billionaire entrepreneur Sir Richard Branson, a certain cachet is not surprising. Nor does it hurt that Virgin Galactic has partnered with Scaled Composites, whose team won the ten million dollar Ansari X Prize, in 2004, for private suborbital spaceflight.[17] (I'll have more to say below about space-related prizes.) Better even than credentials, Virgin Galactic has *customers*: three hundred reservations and forty million dollars in deposits. A ticket currently costs two hundred thousand dollars; for a mere twenty thousand dollar deposit, you can hold a place in line. Forecasted start of operations: early 2012.[18]

(Update: In October 2011, Virgin Galactic reforecast the start of commercial operations as sometime in 2013.[19])

---

16. http://www.spacetoday.org/Astronauts/SpaceTourists.html

17. http://space.xprize.org/ansari-x-prize

18. http://www.crunchgear.com/2010/09/27/virgins-richard-branson-commits-to-space-tourism-within-next-18-months/

19. http://online.wsj.com/article/SB10001424052970204777904576653690338241146.html

What will you get for your money? A trip that soars above the Kármán line,[20] roughly six minutes of weightlessness,[21] and one heck of a view.

Nor is Virgin Galactic the only company pursuing the space tourist trade. Other contenders include Blue Origin (backed by Jeff Bezos, founder of Amazon), Armadillo Aerospace (founded by John Carmack, a cofounder of gaming company id Software), and XCOR Aerospace.

The big aerospace companies are getting interested, too. There's the EADS Astrium Space Tourism Project, backed by the European Aeronautic Defence and Space Company (best know for its Airbus division).[22] And Boeing recently announced a partnership with Space Adventures to offer a "space taxi" service to the ISS by 2016.[23]

---

20. The Kármán line (after physicist Theodore von Kármán): one definition of the border between space and Earth's atmosphere. This conceptual boundary occurs at an altitude of about 100 kilometers, where the air is so thin that the speed needed to achieve aerodynamic lift exceeds orbital velocity. Cross the Kármán line and *voilà*, you are (by NASA standards) an astronaut.

    Hence the 100 kilometer (about 62 miles) altitude threshold set for the Ansari X prize.

21. Six minutes spent weightless in a flight of about two and a half hours duration. The trip begins as a plane ride, with the spacecraft (SpaceShipTwo) tucked beneath the wing of a mother (air)ship (WhiteKnightTwo). The spacecraft drops free at an altitude of about fifteen kilometers, soars for a while on rocket power, coasts to its peak altitude, then glides back to Earth.

    And those vehicle names? One wonders why a space venture can't afford a few more spaces.

22. http://www.space-tourism.ws/EADS-Astrium.html

23. http://www.washingtonpost.com/wp-dyn/content/article/2010/09/15/AR2010091506375.html

### *Spaceports*

When space tourism really takes off (no, I'm not really sorry), where do you begin? It appears that there will be several departure points from the U.S, to include:

- Spaceport America, in the New Mexico desert, billed as "the world's first purpose-built commercial spaceport."[24]
- Mojave Air and Space Port, in California, billed as "the first civilian spaceport," and from which SpaceShipOne made its X Prize-winning flight.[25]
- For those of you nearer the East Coast, there's the Mid-Atlantic Regional Spaceport (MARS, get it?), built on land leased from NASA's Wallops Island (Virginia) facility.[26]

### *All suited up, and no place to go*

Suborbital flights are inherently short duration. What about flights to Earth orbit?

The ISS has a design capacity of six people. If every berth aboard were to be offered to tourists (hardly practical, since the ISS requires ongoing maintenance), that still wouldn't support much business. We need more facilities in space—and it appears that we're going to have them.

Bigelow Aerospace has already launched two generations of prototype orbital habitats. Founder and president Robert T. Bigelow knows something about building lodgings—he owns hotel chain Budget Suites of America. Bigelow Aerospace forecasts that their first operational space station will open for commercial use in 2015. That first space station is aimed at government and corporate

24. http://www.spaceportamerica.com/

25. http://www.mojaveairport.com/

26. http://www.marsspaceport.com/

customers, not tourists.[27]

Russian aerospace firms Orbital Technologies and Energia have announced plans to open a commercial space station by sometime in 2016.[28] Targeted customers for this facility include "private citizens." I read that as "space tourists."

And a dropout along the way? In 1999, Hilton International announced its Space Islands program, an orbital facility to be constructed from and within spent space-shuttle external tanks.[29] Of which none exist in orbit: NASA intentionally splashes the tanks into the ocean rather than clutter near-Earth space with them. Space Islands was a nonstarter.

### *Be prepared*

Space is an unforgiving environment, not someplace to visit unprepared. Also, losing customers makes for bad PR. So: your trip to space—even a suborbital daytrip—will likely include a preflight physical exam, a stint in a centrifuge to confirm your ability to withstand the forces of takeoff and reentry, and zero-gee training.[30]

For longer flights and orbital sleepovers, more intensive training will be in order. Pioneering space tourist Dennis Tito reports of his preflight preparations that, "The comprehensive training program I completed closely mirrored that of any Russian cosmonaut engineer." Tito trained at Star City, the Russian astronautics facility near Moscow. Space Adventures offers a variety of related opportunities at Star

---

27. Early reports about Bigelow's venture—and ongoing rumor—say otherwise, as in this 2005 article: http://www.popsci.com/scitech/article/2005-03/five-billion-star-hotel.

28. http://www.space.com/businesstechnology/russia-space-tourism-commercial-space-station-100929.html

29. http://news.bbc.co.uk/2/hi/sci/tech/293366.stm

30. Virgin Galactic specifies a three-day program of "preparation, bonding and training" for passengers of its upcoming suborbital flights. See http://www.virgingalactic.com/overview/training/.

City, whether or not you intend to leave the ground, including Soyuz spacecraft and ISS training simulators, spacesuit training, and (for the scuba-certified) spacewalk training in a neutral buoyancy tank.

Nor need you travel to Russia or book a spaceflight to get astronaut training experience. Aurora Aerospace, in Florida, offers what one reporter called "space camp on steroids."[31]

### *Prizes*

Conventional wisdom has it that prizes spurred progress in early aviation. Perhaps the most famous aviation contest was for the $25,000 Orteig Prize, offered in 1919 by hotelier Raymond Orteig, for the first solo trans-Atlantic airplane crossing. Charles Lindbergh won the Orteig Prize in 1927.

The X Prize Foundation was established in 1996, expressly citing the Orteig Prize as their inspiration, to extend award incentives to the development of nongovernmental spaceflight. A dollar doesn't go as far as it did in 1919 or 1927; the minimum X Prize is for $10 million.

I've already mentioned the Ansari X Prize, awarded in 2004 for demonstrating a reusable, privately funded, crewed, suborbital spacecraft. This wasn't the foundation's only challenge. Their biggest contest is (no points for guessing the identity of its chief corporate sponsor) the Google Lunar X Prize. To collect $30 million, all you need to do—before the end of 2015, and using only private funds—is land a mobile robot on the moon, send it five hundred meters across the lunar surface, and transmit back video, images, and data.

And do it before anyone else. Twenty-two teams have registered.

NASA has its own space-centric, prize-paying challenges aimed at inspiring/obtaining new technologies.[32] To name a few:

---

31. http://www.space.com/businesstechnology/090602-astro-training.html

32. www.nasa.gov/challenges

- The night-rover challenge, for $1.5 million, deals with solar-powered robots able to store energy for nighttime explorations.
- The power-beaming challenge (partially awarded, $1.1 million remaining) seeks technology for wireless power transmission. Merely use beamed power to drive your robot one kilometer up a vertical cable.
- The sample-return challenge, also for $1.5 million, to demonstrate the ability to locate and return samples from varied terrain.

### *Climbing out on a limb*

That's the past, present, and very near future of space commercialization.

What else might we expect? This is not the place to be shy about prediction: science fiction has long anticipated commercial activities in space. (A classic example: Robert Heinlein's 1950 novella "The Man Who Sold the Moon"). And who among SF authors has not at least once contacted Central Casting for a doughty, stubbornly independent Asteroid Belt prospector?

For the remainder of this article, we'll look at longer-term and more speculative possibilities than tourism and cargo delivery.

### *Powersats*

Sooner or later—whether because oil becomes too costly or too difficult to extract, or to reduce humanity's carbon footprint—mass transportation will switch from internal combustion engines to all-electric cars. We're talking about a *lot* of cars: call it one billion worldwide, about a fourth of them in the United States.[33]

That switchover can't proceed to any meaningful degree absent a major expansion in electrical power capacity. (A scaled-up electrical

33. http://www.plunkettresearch.com/Industries/AutomobilesTrucks/AutomobileTrends/tabid/89/Default.aspx

power grid is but one obstacle to the switchover to all-electric cars. Other factors—like improved battery designs and the scarcity of materials like lithium to manufacture those batteries—are beyond the scope of this article.) Solar energy, renewable and pollution-free, is expected to drive a large part of the expanded power generation capacity.

Alas, sunlight gathered on the ground is less than ideal as an energy source. The sun isn't helpful at night, and even by day clouds and precipitation impede collection. Atmosphere scatters sunlight—as in: why is the sky blue?—before much sunlight can even reach ground-based solar cells. The most effective sites for solar farms (e.g., southwestern deserts) aren't the places that need power (e.g., New England). Collection, overnight storage, and distribution of solar energy will require a lot of land—so-called energy sprawl.

In contrast, solar cells in geosynchronous Earth orbit (GEO) collect sunlight effectively 24/7,[34] undiminished by any atmosphere. That solar energy, converted to microwaves or laser beams, can be sent to the ground at frequencies to which the atmosphere is effectively transparent. If the downlink beam is steerable, power can be sent directly to the region where it is most valuable—perhaps Miami on summer afternoons and Boston on winter nights. The downlink beam would be more intense than incident sunlight, so collection stations can be smaller than a terrestrial solar farm of equivalent capacity.

Why has no one built a powersat? In a word: mass. As one possible implementation, consider the powersat reference model defined by the International Union for Radio Science.[35] To feed a gigawatt

34. The brief exceptions occur at night, around the equinoxes. These are times when power demands are comparatively low. See article about geosynchronous orbits at http://webphysics.iupui.edu/gpnew/gp1sup2b.htm.

35. Officially, the Union Radio-Scientifique Internationale (URSI). See "URSI White Paper on Solar Power Satellite (SPS) Systems," at http://ursi.ca/SPS-2006sept.pdf.

of power[36] into a terrestrial electrical grid the model begins with a ten-square-kilometer satellite that masses over a million kilograms (a thousand metric tons). For reference, that's three times the mass of the ISS—which has taken more than a decade to assemble. Merely delivering the components of the powersat to orbit would cost billions of dollars.

(The CRS program's derivable cost per metric ton includes some recovery of R & D and launch system construction costs. Additional deliveries might therefore be priced significantly lower. Economies of scale should kick in, given the many flights that would be needed to transport into orbit the parts for even one large powersat.)

But there's another wrinkle: the proposed space-taxi services are meant to make deliveries to LEO, a few hundred miles above the Earth, not GEO, about twenty-two thousand miles above. It's often assumed, as I have in my novel *Energized*, that powersats would be assembled in LEO and slowly boosted to their final orbits. After all, powersats generate lots of electrical power that can power onboard ion thrusters.

All that mass (delivery cost) is a challenging problem, but perhaps not insurmountable. California power utility Pacific Gas & Electric has announced plans to buy space solar power from Solaren Corporation starting in 2016.[37]

36. A gigawatt (billion watts) only sounds like a lot. By 2008, the U.S. capacity for electrical power generation exceeded a terawatt (trillion watts). In 2006, worldwide generating capacity was about 2.75 terawatts. (See details from the U.S. Energy Information Administration at http://web.archive.org/web/20101203180131/http://www.eia.doe.gov/cneaf/electricity/epa/epat1p2.html and http://www.eia.doe.gov/pub/international/iea2006/table64.xls.)

37. "Space-Based Solar Power Coming to California in 2016," Physorg.com, April 15, 2009, http://www.physorg.com/news159020477.html.

### The moon

We now know the moon holds vast amounts of water ice.[38] Ice means water to drink, oxygen to breathe, and hydrogen to fuel rockets. Lunar colonization suddenly seems possible, and some of the organizations interested in commercial operations near Earth have also expressed interest in lunar projects.[39]

NASA has shelved plans for a return to the moon, but Japan, China, and India are in something of a second moon race.[40] And why, other than for prestige and out of scientific curiosity, might anyone care to revisit the moon? Perhaps, for energy. The lunar surface has for eons absorbed helium-3—both a fusion byproduct and a prospective fusion fuel—from the solar wind. Apollo astronaut and geologist (and former U.S. senator) Harrison Schmitt estimates that lunar He-3 collected as fuel for terrestrial fusion power plants could be competitive with oil at $30 per barrel (vs. about $80 per barrel around the time I write this).[41] If only we had controlled fusion …

And perhaps the moon is for tourism, too. Space Adventures—without providing much in the way of specifics—is shopping for customers for a flight around the moon.[42]

### Asteroids, near and far

Asteroid mining is a science-fictional staple for good reason: *lots* of valuable metals are locked up in asteroids. Studies on the subject

---

38. http://www.universetoday.com/58410/water-ice-found-on-moons-north-pole/

39. http://www.space.com/businesstechnology/moon_conference_000724.html (Shown for historical purposes only. This article has vanished without a trace.)

40. http://www.atimes.com/atimes/South_Asia/IH29Df01.html

41. *Ibid.*

42. http://www.spaceadventures.com/index.cfm?fuseaction=Lunar.Details

bandy about dollar figures in the billions and trillions. As in, from a BBC report on the asteroid Eros, "The most detailed study of an asteroid shows that it contains precious metals worth at least $20,000bn."[43] That's on a single asteroid! Other asteroids and comets offer materials far more mundane—yet essential to human life—like oxygen, water, and various carbon compounds. Those are valuable, too, eliminating the need/expense to haul them out of Earth's gravity well.

"All" we need is improved spacecraft to take us to and from the asteroids. Might we get such craft? Relieved of LEO cargo duties, NASA's next big effort (according to its recent budget blueprint), will be a crewed asteroid mission by 2025.[44]

The Asteroid Belt lies well beyond Mars, but sometimes asteroids and comets come to us. When a Near Earth Object smacks Earth, that's a bad thing. Ask any dinosaur. (Oh, wait.) But if an asteroid does head our way—and thousands come close regularly—we could capture it into Earth orbit.[45] Capture a NEO or two, and (again, as in my novel *Energized*), the onsite building material will make powersats much more affordable.

Whether into orbit or far away, humanity needs to be prepared to deflect an oncoming space rock.

(December 2011 update: Asteroid 2005 YU55, 400 meters across, sped by on November 8, 2011. It passed closer to Earth than does the moon. Without doubt, that was a Potentially Hazard Object—and still beyond humankind's ability to divert or capture.)

---

43. "Gold Rush in Space," BBC News Online Science Editor Dr David Whitehouse, July 22, 1999, http://newsimg.bbc.co.uk/1/hi/sci/tech/401227.stm.

44. "NASA in Transition as Congress OKs New Direction," *op. cit.*

45. See "Rock! Bye-bye, Baby," elsewhere in this collection.

### *Helping "hands" and armchair tourism*

Companies use robots for tasks that are repetitive, dangerous, or take place in environments that are inhospitable or inaccessible. Space is certainly an inhospitable environment, and many operations in space—such as prospecting, mining, and manufacturing—may prove both dangerous and repetitive. Forecasting robot workers in space doesn't seem like a big leap.

But I don't mean thinking, self-aware electromechanical colleagues, because artificial intelligence remains A Very Hard Problem. That's okay: much can be accomplished with today's robotic technology. Remotely operated vehicles routinely service underwater oilfields and further demonstrated their utility after the 2010 Gulf of Mexico oil-well blowout. With appropriate means of locomotion, remotely operated robots will serve humanity in space, too.

Rather than AI, the key technology is telepresence: the combination of remote control and immersive sensing (aka, virtual reality). Telepresence technologies being developed for remote surgery, to make scarce specialists more widely accessible, will also facilitate remote mining. At first such robots' human controllers may be nearby, as in NASA's trial of a robot worker aboard the ISS.[46] But, like deep-marine ROVs, some space bots will be operated long-distance. JAXA, the Japanese space agency, plans a robotic lunar base built by robots, for robots.[47]

Care to try telepresence today? Startup firm Anybot is taking orders, and beta-test units have already been seen in the (terrestrial) workplace.[48] While the first bot on the ISS is humanoid, the Anybot

---

46. "NASA Ready to Send Humanoid Robot to Space," by Courtney Boyd Myers, *IEEE Spectrum*, September 3, 2010, http://spectrum.ieee.org/automaton/robotics/humanoids/nasa-ready-to-send-humanoid-robot-to-space.

47. http://www.techngadgets.com/japan-plans-a-moon-base-by-2020-built-by-robots-for-robots.html

48. "When My Avatar Went to Work," Erico Guizzo, *IEEE Spectrum*, September 2010, http://spectrum.ieee.org/robotics/industrial-robots/when-my-avatar-went-to-work/.

QB robot looks more like a floor lamp on wheels. As for "hands" in the section heading, a robot's grippers may differ from human hands as greatly as wheels differ from feet. Snakebots, anyone?[49]

It'll be a short step from telepresence via space robots to armchair space tourism. This once, I can't find a company preparing to offer the service, but that's okay—I'm a card-carrying SF author and futurist. I'm confident that within twenty years I'll be able to rent telepresence by the hour on lunar or NEO robots. And it'll be a lot cheaper than flying to the moon.

Robots might also do the monotonous and, perhaps, dangerous work of fabricating structural elements and electronics (say, solar cells and microwave transmitters) from the on-orbit resources of a captured NEO. Robots can assemble such components into complex orbital structures. Powersats are looking better and better.

Telepresence has its limits—at least for impatient armchair tourists. The roundtrip radio delay between my armchair and the moon is about 2.5 seconds, a bit longer if the signal is relayed through comsats orbiting the Earth or moon. Building in an execution delay of, say, ten seconds between receipt of command and start of movement would provide the opportunity to override many ill-chosen commands. I wouldn't want to send my rent-a-bot tumbling off a lunar cliff (at least, I wouldn't want to forfeit my deposit). Alas, building in safety delays becomes cumbersome for other places I might like to visit telepresently (to coin a phrase). For example, Mars—at its closest approach—remains a bit over four light-minutes from Earth.

### *Enter the lawyers*

Plucky bands of asteroid prospectors presuppose property rights (repeat that quickly ten times) to whatever treasure might be found in the Belt. Can an individual or a company have property rights in space?

It's not clear.

---

49. http://www.techngadgets.com/the-army%E2%80%99s-new-robotic-tentacle-manipulator-uses-teams-of-snakebots-to-manipulate-objects.html.

What *is* clear is that the Outer Space Treaty (officially, the Treaty on Principles Governing the Activities of States in the Exploration and Use of Outer Space, including the Moon and Other Celestial Bodies) declares that celestial bodies are the common heritage of all mankind. (And if intelligent aliens exist to contest our claims? We'll skip that complication.) A country can't legally claim a celestial body. You or I certainly can't own one.[50]

Still, as everyone knows who has ever rented an apartment or leased a car, one can separate the right to use property from ownership of that property. Humanity, as the self-appointed collective owner of celestial bodies, can assign property rights—like the right to mine an asteroid for metal, or the moon for He-3—to a country, company, or even an individual.

Will humanity, presumably through the aegis of the United Nations, find a way to sell, lease, or grant valuable off-world property rights? As an encouraging precedent, consider the International Telecommunications Union, the United Nations agency best known for overseeing allocation of radio spectrum worldwide. The ITU also allocates scarce GEO slots, primarily used by comsats. The allocation process is often slow and cumbersome—but it works. Countries do obtain the rights to use GEO slots, and some countries in turn have delegated their rights to commercial entities. And so, DirecTV's many million subscribers get their television shows.

Other precedents are less encouraging with respect to utilizing "common heritage" assets. Many ocean resources—like all space resources—are in an inhospitable, hard to reach, expensive to exploit, international environment. The international Law of the Sea Treaty (of which the United States is not presently a signatory) governs the oceans—and rules for mining in international waters have yet to emerge. In another example, consider Antarctica, over which sovereignty claims are few, conflicting, not universally recognized, and

50. "Space junkies ask: 'who owns the moon?' ", Stefanie Olsen, CNET, August 23, 2008, http://news.cnet.com/8301-11386_3-10024047-76.html.

constrained by other international treaties. The Madrid Protocol to the international Antarctica Treaty bars mining on the south-polar continent until 2041.

(The case can be made that some places should be preserved or protected rather than developed. I won't make that case here, because nature conservancy is outside the province of this article's focus: commercial development.)

The time-honored anarchy of the frontier suggests another possibility. If you illegally obtain a "common heritage" resource out in the lunar hinterlands or deep in the Asteroid Belt, who (besides, extending the Wild West analogy, rustlers and claim jumpers) can stop you?

Will anyone even know? Gold and platinum are routinely valued above a thousand dollars per ounce. And as we've discussed, *any* useful material—say, lunar ice—delivered to LEO inherits the value of foregone boosting-from-Earth costs. Refine precious metals from an asteroid or purify lunar-polar ice, and who is to say where the stuff came from?

Suppose property rights in celestial objects become established. Who will have jurisdiction and authority to enforce them? That's also less than clear. Until 1989, when the U.S. Marshals established an office, Antarctica had no law-enforcement presence.[51] Legal jurisdiction in the American-administered section of the continent seems straightforward—but what jurisdiction can the marshals exercise in parts of the continent administered by another country?

When will exploitation rights to bodies in space, and law-enforcement mechanisms to safeguard those rights, begin to be clarified? I predict: not until humanity's "common heritage" begins to experience serious pilferage.

### *Wrapping up*

Will the private sector take up the torch that NASA hopes to pass? Will a private-public partnership reinvigorate American space exploration? Will space become the new frontier of commerce and industry?

---

51. http://www.usmarshals.gov/history/antarctica/

It could all happen. The hopeful portent is that companies—often with billionaire private backers—are, as we've seen, investing serious money to chase opportunities in space.

While the future of space exploration sorts itself out, my advice remains: Follow the money.

### *Time flies (like a rocket)*

Between completing this article (October 2010) and receiving galleys to review (May 2011), commercial spaceflight kept making progress. A few news items:

December 2010: SpaceX orbited and recovered a large cargo capsule—the first demonstration flight under NASA's COTS program.

February 2011: Southwest Research Institute executed contracts with Virgin Galactic and XCOR Aerospace for scientists to do research aboard both companies' suborbital spacecraft.

April 2011: SpaceX announced plans for a heavy-lift booster (the Falcon Heavy) with twice the payload capacity of the retiring NASA space shuttle. Targeted first flight: 2012.

### *Make that: a really fast rocket*

Another few months passed before I began assembling this collection—and so, it was time for another update. The world of commercialized space travel changes quickly, and not always for the better. A few recent news items:

Congress scaled back NASA's fiscal 2012 budget request for bootstrapping commercial-space operations: The $406 million authorized is less than half the $850 million requested for CCDev, the commercial crew program.[52] NASA, as a consequence, has delayed its purchase of commercial crew-carrying flights to the ISS until 2017 at the earliest.[53]

---

52. http://www.washingtonpost.com/national/health-science/nasa-and-national-science-foundation-are-spared-from-big-budget-cuts/2011/11/16/gIQAz7FwhN_story.html

53. http://www.space.com/13953-nasa-commercial-crew-space-act-agreements.html

The first private resupply mission to the ISS under NASA's CRS program, to be flown by SpaceX, was first targeted for late 2011. The launch slipped to January 2012 and most recently to, at the earliest, March 2012.[54]

Blue Origin, one of NASA's CCDev program contractors, lost a rocket on an early test flight.[55]

Despite setbacks, the players evidently continue to see a bright future in commercialized space operations. SpaceX is scouting out sites for a "private Cape Canaveral" for its operations.[56] Boeing has made arrangements to prep its CST-100 "space taxi" at Cape Canaveral, in the space-shuttle program's now disused hangar.[57] Virgin Galactic has opened new offices in Los Cruces, New Mexico, home of Spaceport America.[58] Billionaire Microsoft cofounder Paul Allen unveiled Stratolaunch Systems—yet another private space-launch company—aspiring to private orbital flights within five years.[59]

---

54. http://news.cnet.com/8301-19514_3-57363144-239/testing-pushes-spacex-cargo-flight-to-at-least-late-march/

55. http://www.ibtimes.com/articles/208015/20110903/jeff-bezos-highly-secretive-blue-origin-test-vehicle-crashes-amazon-chief-boeing-nasa-commercial-cre.htm

56. http://latimesblogs.latimes.com/money_co/2011/11/spacex-new-launch-site.html

57. http://www.nasa.gov/exploration/commercial/OPF_Boeing.html

58. http://www.bizjournals.com/albuquerque/news/2011/11/28/virgin-galactic-leases-las-cruces.html

59. http://www.space.com/13915-paul-allen-giant-plane-stratolaunch-systems-private-space.html

***To read further:***

***Space news, including commercial space ventures:***

http://www.space.com/
http://www.spacedaily.com/

***Space tourism:***

http://www.spacefuture.com/tourism/tourism.shtml

***Using near Earth asteroids and lunar material:***

http://www.permanent.com/a-index.htm
http://www.permanent.com/l-index.htm

# *Small Business*

The blue shirt moved out of the holo at a glacial pace, as though sliding to the left down a very sticky inclined plane—not that, in microgee, any plane inclined more than any another. Just as a hint of the objective finally appeared, the bored technician, to the soft *zzzp* noise of Velcro parting, sidled in front of the camera. Sweat-stained denim once again filled the scene.

There went an hour's slow progress.

"Damn," Jason Grimaldi muttered. He batted aside a floating clipboard. His companions shushed him, although the only sound accompanying the video at that moment was the *whirrr* of a ventilation fan. Jason studied the wrinkled garment for clues to the nearest side of the obstruction.

"Should we wait? He might step back." The comment was a vintage Bill-ism. Jason did not get how someone who always hoped for the best became a revolutionary.

Jason nudged a joystick. "Too risky, with the batteries so low. I won't make a bet on how he'll next move, either. I'm going for altitude." Most rooms in the space station had an orientation, even when nothing but convention distinguished floor from walls from ceiling.

The view in the holo display crept upward. A frayed collar came into view, then a neck. The nightshift stuckee needed a new uniform and a haircut. Channeling Bill, Jason hoped the sloppiness denoted lax discipline. They needed all the help they could get.

Moving to either side of the neck would bring their objective into view the fastest—unless the tech moved again. Checking the power readout, Jason rejected the gamble. The gauge was approaching redline; they might not recover from another setback. So: onward and upward.

It would be minutes until they could see over the man's head. Jason stretched tense muscles as best he could. At two meters and almost a hundred kilos, he overflowed the standard-sized command seat. The bridge, alas, offered no room for pacing.

In the holo, salt-and-pepper hair gave way to shiny scalp. The target peeked at them through sparse wisps.

The batteries were down to two percent.

"Increase the focal length," Sherry suggested.

The hair in the foreground softened to a blur; details of equipment across the room became maddeningly *almost* discernible. Jason panned. The image vibrated as though, as though … As though what? The camera platform's steady pace was not the problem—computers compensated for that. Jason gnawed on a pencil stub, staring at the fuzzy picture. Vibration? Why would there be …?

Uh-oh. He cranked up the volume of the audio pickup. What was that tone beneath the ubiquitous fan noise? The microphone must be very near an unsuspected air vent to capture the bass rumbling of a duct. *Too* near.

Target and technician alike spun from sight as Jason lunged for the joystick. For a time, the image tumbled too quickly for digital correction. The scene finally resolved into an ant's-eye view of the console shelf onto which air currents had delivered the 'bot. The speaker emitted *zzzps* Jason had no problem identifying: approaching footsteps. Over the tearing sounds came a disgusted voice: "I *hate* bugs."

They glumly watched the descent of a bent-double sheaf of printout.

◆❖◆

Six months earlier, Jason would never have expected to end up a revolutionary either. Comfortably settled at K-State, he pursued his studies with monomania and a few close friends. When the mood was right, he and his university buddies might rail against the injustices of life. Why not? Griping was free, and a way to let off steam.

Still, Jason listened more than he griped. His parents drove FedEx trucks. No matter what those boxes contained, delivery could only happen locally. He gave little thought to where the goods in those boxes were made. Preoccupied with his dissertation research, his material needs were few. His stipend as a graduate assistant paid the rent, and he tutored for mad money.

That complacency vanished the day Sherry Nilsson appeared in the Student Union. Every move she made declared her a spacer. Somehow she made the unaccustomed struggle against gravity graceful and exotic. He admired her from afar, giving no thought to the improbability of a spacer picking Manhattan, Kansas as a destination. Looking back, he didn't think much in those days—or the thinking he did wasn't done with his head.

Opposites truly did attract. Sherry was as willowy as Jason was burly, as blond and Nordic as he was dark and Mediterranean, and—the biggest contrast of all—an econ postdoc. Had he found buttons cute, doubtless he would have added cute-as-one to Sherry's list of virtues. She drew Jason like a moth to a flame, and he could only hope that his fate would be less clichéd and fatal.

It would be weeks before he wondered whether Sherry was here precisely because she was his "type." Or that his technical skills and working-class roots, not his diamond-in-the-rough charms, had attracted anyone's interest.

Jason tried to insinuate himself into her circle. Sherry and her friends seemed always to be debating politics or economics. He often couldn't tell which. He decided that the boundary wasn't necessarily clear-cut.

With uncharacteristic patience he dallied at the outer fringes of her clique. Then one day the main topic was government regulation of innovation. He swallowed hard. The verbal jousting had at last touched on a subject in which Jason could claim expertise. It had something to do with technology, anyway.

It was time to make his move.

Jason coughed softly. He cleared his throat. He eventually got

Sherry's attention, which turned out to be as unsettling as her smashing good looks.

"So what's the big deal about these little machines?" Sherry asked when he named his field, smiling at her own pun. Aimed directly at him for the first time, her ice-blue eyes almost left him tongue-tied. "Of what conceivable use is a millimeter-long gear or motor? And if smaller is better, why settle for half-measures? Why not go all the way to nanotech?"

He rallied: This was his chance to make an impression. "Of what use is *any* gear or motor? We size machines for the tasks they handle. Micromachines ideally suit specific tasks."

"Surgery on mosquitoes?"

He nearly melted at the saucy curl of her lip even as he took exception. *Lord, but he was smitten.* "No, Sherry, although I could probably build the tools for that job."

Micro electromechanical systems were a narrow specialty even within the engineering field. MEMS robotics was a niche indeed. K-State had a solid program, as did a handful of other schools worldwide, but that was it. Jason had plenty of experience explaining his passion to relatives and friends—and that MEMS devices were about a million times larger than nanobots.

He chose an example that worked for most laypeople. "Look, say an optical fiber goes dead in a trans-Atlantic cable. Signal repeaters are hundreds of kilometers apart, so a break—say, from a shark bite—could be anywhere between. Locating the damage by timing reflections from the break isn't precise. Searching for a break by sub or remote-controlled submersible is damned expensive. So is laying new cable. *That* is a suitable problem for micromachines."

"Huh." She cocked her head. "What does a micromachine do? The backstroke?"

He had to laugh. "Imagine a gnat-sized robot able to move among the fibers within the cable sheath. Deploy a few gnatbots in every repeater. If a repeater loses signal from one direction, it sends a gnatbot creeping down the cable toward its silent neighbor. When the 'bot sees light leaking out of a damaged fiber, it's found the problem. Having counted its steps along the way, it reports back the exact location of the break."

"How?"

Her friends looked bored. Some wandered off. For all their haranguing about technology regulation, few of them ever showed *interest* in technology.

"How can it report?" Jason grinned. "It's easy. Look, the robot is tiny. It's inside a fiber-optical cable. It simply polishes the broken fiber end into a usable state and beams its own light signal down the fiber." It could even be designed with a bit of fiber running through its body. Polish *both* ends of the break, and splice the ends to itself: repair the quick-and-easy way.

Sherry considered. "That's clever. But how would you control such a gadget?"

He reveled in her interest. "For a task that simple, I'd program it to be autonomous. For more complicated jobs, adding a low-power laser comm link is appropriate."

One of the hangers-on inched closer. He was a jaundiced, gaunt fellow, a painter of some kind enamored of the starving-artist tradition. "Could your creature find its way back out?"

"Device," Jason corrected automatically. "Sure, but why bother? Gnatbots should be dirt cheap." And for that reason, the 'bots would not be stationed hundreds of kilometers apart, only in the repeaters, but scattered along the length of the cable.

People looked surprised at that answer. (Jason did not process at the time that *Sherry* showed no surprise. It all came clear later.)

Jason said, "A micromachine is a few pennies-worth of silicon and metal, no different than an integrated circuit. The machinery and its controls are formed at the same time, by the same industrial processes, from the same chip. The big costs are upfront: design labor and manufacturing equipment." Like nanotech, for that matter.

More of Jason's audience excused themselves, headed for a laser-art exhibition. His ire—he'd listened to their preaching often enough—vanished when Sherry made no move to follow.

"So you *can* talk. I had begun to wonder." She laid a hand on his elbow. “Buy a girl a cup of coffee, sport?”

◆❖◆

For weeks, he and Sherry discussed politics, life in space, economics and—to his unending surprise—microengineering. Some of her questions, hindsight being 20/20, suggested more than passing familiarity with MEMS.

Sherry’s friends kept happening by to chime in. Over beer and pizza one evening, when they were curiously alone, Jason asked when their chaperones were due.

“I deserve that.” Sherry sighed. “You really understand microengineering, don’t you?”

Some sixth sense told him that, contrary to appearances, she had not changed the subject. “No offense, but how would an economist know?”

“Does Ron understand poli sci?”

Ron was one of Sherry’s maddening crowd. Ron sounded off now in Jason’s mind’s ear, pedantic and domineering. Jason grimaced. “I have my doubts.”

“Why?” Sherry persisted.

“Ron is too sure of himself. Too much of a know-it-all. I can’t respect someone incapable of saying ‘I don’t know.’ ”

She smiled. “While you pick words carefully and consider the ramifications of what you say. And you do occasionally say ‘I don’t know.’ *That’s* how I’m sure you know your stuff.”

By that standard, he would concede she knew economics. “What bearing does this have on your chaperones?”

“Co-conspirators,” Sherry whispered, standing. “We were evaluating you.” And in the course of a long, moonlit stroll, she explained.

◆❖◆

“Are you with us?”

It was obvious in retrospect: Her cronies had been probing Jason for his political views. Sherry was the bait to keep him there. It was unclear whom that admission most embarrassed.

What did he think? With a few reservations, he sympathized with her aims—and so what? The ends were not the issue. The proposed means were.

And prison, quite likely, if things went wrong.

Had Sherry educated him? Indoctrinated him? Were hormones making his decisions? Or did they simply agree?

"Jason?" Sherry's voice cracked, her face drawn with worry.

Every prospective recruit was a prospective informant or undercover Syndicate agent. She had not decided lightly to approach him.

This was not, Jason decided, a deceitful face. Yes, he was in—although he had a final bit of due diligence to perform. He would do that on his own.

Whatever he learned, he wouldn't—couldn't—turn in Sherry. "Okay, I'm in."

She slumped in relief.

He continued, "Sure, I could use the university microengineering labs to build you tiny spybots. I could teach you and your friends to control them.

"But I have a better idea."

◆❖◆

"... Announced today that the *SSS Helsinki* will be commissioned on Tuesday. The *Helsinki*, eighth and newest member of the Syndicate's deep-space fleet, will accommodate missions of up to a year's duration. Equipped with the latest engines, the *Helsinki* will be the fastest of the Europa-class cruisers. Syndicate spokesman Alain Lamoureaux stated that ..."

Shortness of breath did nothing to mitigate the propaganda. Freefall jogging only looked easy. Every step Jason took on the treadmill tried to bounce him—equal and opposite reactions made undeniable—into the air. Powerful bungee cords hooked to belt loops yanked him back. Wrist and ankle weights (okay, not *weight*, but still mass: resistance in the form of added inertia to be fought) intensified the workout.

*Exercise here. Be pushed from the ship in a wheelchair when we return to Earth.* It's your choice, Sherry had said.

So he exercised every chance he got.

Just twenty minutes offworld Jason had hurled, spraying lunch and probably the two meals before that. Loudly and sloppily. And floated through his own drifting vomit, too nauseous to care. That was plenty to live down.

His stomach lurched at the memory, and he tried to focus on the newscast. Four Europa-class ships were being shown. At the center of the image utility craft swarmed. They serviced the nearly outfitted *Helsinki*, hanging in space beside its birthplace: Syndicate Station Three. The camera shot came from the tethered sensor pod that orbited three hundred meters Earthward from the main body of the station.

Jason trudged on, his sweaty tunic clinging. The few metal ships looked crude and quaint among the translucent diamond hulls of the nano-grown Syndicate ships. Craning for a different view into the holo, he glimpsed the metal ship docked at Beta port.

This ship.

For years, the *Tom Paine* had shuttled between Earth and the Lagrange points, tapping the mineral wealth of the Earth/Luna Trojan asteroids. Scarcely a third the length of the sparkling Syndicate cruisers, the *Tom Paine* was representative of independents' ships: an antique.

The logic was inescapable. Metal hulls massed more than carbon, burning more fuel. Metal hulls were less dependable. A few grams of nannies kept the new-style hulls in repair—not that much could harm a diamond—but a metal ship needed an inventory of spare parts. Spares sacrificed that much more potential cargo capacity. And lugging extra fuel and spare parts required even more fuel …

When diamond spaceships began working the Trojans, independent miners could not compete.

Independents like Bill Nilsson remained in space any way they could. Too often that meant demeaning support contracts for near-Earth Syndicate operations, and even that work was disappearing. They saved their pennies and dreamt of the day when they would buy their own synthetic diamond hulls—

Until, Bill said, even the cockeyed optimists had to accept the truth. The Syndicate would never, for any price, equip potential rivals.

Breathing deeply to clear his head, Jason tried again to focus on the vid.

"In other news today," the 'caster went on, "Edouard Smithson, France's ambassador to the United Nations, formally introduced the much-discussed global legislation to ban nanoengineering activities on Earth. Citing two recent loss-of-containment incidents on Syndicate Station One, the ambassador called nanites, 'Too great a risk to life for prudent terrestrial development.' "

The holo cut to Smithson, a long-faced man with eyebrows like woolly caterpillars. In dolorous tones, the ambassador said, "While my colleagues and I recognize the promise of nano-fabrication, we must not—and we will not—endanger life on Earth through hasty and potentially irreversible experiments. The technology can only be developed safely in space, where any incidents are intrinsically isolated."

Jason managed, just barely, not to shout at the vid. How interesting that these incidents occurred on an obsolete station. Was Syndicate Station One more valuable as an object lesson than as scrap?

Toweling wet hair, Sherry entered the ship's tiny dayroom. "You know, we *do* have exercise videos."

Jason gestured at the vid. "Sanctimonious scumbag. Who besides the Syndicate works with nanites on an industrial scale in space? Everyone else works Dirtside—and without any *incidents*."

She killed the webcast. "We knew this was coming, Jason. That's why you're here. We need to stay on task."

He was *so* sick of this treadmill. The station gym had a dozen kinds of zero-gee exercise equipment, none of which he could use. Docking charges alone were more than they could afford. Every minute any of them spent off the ship added "environmental fees." Breathing was not to be taken for granted in space.

He kept trudging. "You'd never guess from any words leaving

Smithson's lips that the Syndicate is a state-owned enterprise of the Eurasian Union. Or that without nano-fabbed diamond hulls and platforms, no one can compete out here with the Syndicate."

She gave a minimal flick of hand and arm that reminded, "Who told you?"

Syndicate lobbyists were all over this "issue." The money was flowing. Campaign contributions. Economic development aid. R & D grants. Outsourcing contracts. Syndicate scientists normally unavailable for comment were suddenly everywhere on the 'net, gloomily asserting the inevitability of further incidents. Three media giants (all well supported by Syndicate advertising) were promoting epic miniseries about gray-goo disasters. Never mind that, for reasons of safety, no one made nanoconstructors robust enough to survive in the wild. And that people smart enough to build a nanoscale self-replicator were smart enough not to try.

Spacer sympathizers Dirtside predicted Smithson—meaning the Syndicate—had ample votes in the General Assembly to enact the ban.

Sherry's econometric models projected the Syndicate would drive the independents from space within three years, before they could complete their own offworld nano-fabs. The Syndicate would own all the resources of space, with Earth's billions as a captive market.

Jason brushed a sodden lock of hair from his forehead. "Yeah, Sherry. You told me. I guess I didn't *really* believe it. It's easier, somehow, to believe it out here."

He stopped the treadmill. Carefully, he detached the taut bungee cords, then put on his Velcro slippers. They went together to the bridge, and Jason belted himself into the chair at his control console. "I believe it, but I refuse to accept it. Let's go get the bastards."

◆❖◆

Sherry plucked at her hamburger bun, and then in the air at the crumbs she had set loose. Her hair was pulled back in a long ponytail that floated behind her baseball cap. She returned Jason's smile of encouragement.

If only someone could encourage him.

"Very illuminating." Bill spoke from the ship's galley, to which he had disappeared for a coffee bulb. Caffeine megadosing was the older man's one vice. The dour spacer had fallen upon hard times—he was ship's captain, pilot, engineer, and cook.

"Yeah, right." Jason considered another burger, and thought better of it. The first churned unhappily in his stomach. Today's swatted scout was their sixth try at crossing that room.

Bill called out, "No, I mean it. Literally. The lighting in that service area was fairly bright. Could you add photocells to a scout? It could recharge its batteries for the final sprint."

"Hmm." Jason straightened out of a slouch. "If we could …"

They couldn't, of course. He pulled up the schematics anyway, hoping to be wrong.

A gnatbot had six independent legs, each with a dedicated micromotor. Locomotion was the biggest drain on the internal battery that filled the 'bot's guts.

"Zoom, please," Sherry said. He did, and she leaned closer anyway. "Fewer legs?"

Maybe it could get around with fewer legs—but no. Jason shook his head. "Can't, without sacrificing its disguise. To pass for a bug"—unwelcome, but innocent; the station was full of stowaway insects and their spawn—"it needs six legs."

The head was dominated by two integral CCD cameras, each capped with a fisheye lens for 180° visibility. Two piezoelectric sensors astride the head served as ears. Jason thought aloud. "We need both cameras for stereoscopic imaging. We need both mikes to triangulate sound sources."

Four dark regions, stubby steerable rods, marked the robot's back. The right pair were infrared lasers; the left pair photodetectors.

The whole damn surface was in use! Where could he possibly add photocells?

"We're so close," Bill said. "You'll find a way."

Jason closed his eyes. "You know, Bill? That optimism of yours can be really annoying."

The hell of it was they were *so* close.

In just a few days they had stretched a comm network, an invisible daisy chain of two-way infrared links, across the station. (Only one link, from the *Tom Paine* itself to the relay in the docking collar used radio, at less-than-cell-phone power levels, so they could conspire behind closed airlock hatches.) The first relay placements were easy. While playing tourist they stuck 'bots to dimly lit spots in the public corridors. These fixed relays needed only to pivot and twist for aiming.

Then they dumped mobile 'bots into the station's air ducts to extend the network into the guarded and secure Syndicate-only lab annex. Fans did the work, delivering 'bots across the station. The micromachines grabbed hold where they could. In seams between duct segments? To dirt within the pipes? They would never know, and it hardly mattered.

For one 'bot to detect another's flailing laser beam took time, but link by link the network took form. Random distribution meant gaps, but the third 'bot-swarm release plugged the holes. They could see and hear much that transpired in the control room of the Syndicate's main orbital fabrication facility.

Then they sent scouts through the ducts to explore the secret labs.

Scant meters short of their ultimate goal, they were stuck. Repeatedly. For want of a microwatt the gnatbot was lost ….

His sarcasm weighing on his conscience, Jason opened his eyes. "There's hardly any free surface area, Bill. I can't add enough photocells to matter."

Bill made no comment, letting silence speak for him: You're the expert. Make it work.

Jason hummed. He studied the ceiling for a while. He whistled tunelessly. He drummed fingers on the dayroom table. He gave up the tapping to stroke his chin in thought. There *had* to be a way. "Unless …"

Sherry sat up. "Unless what, hotshot?"

"Unless we build the 'bots from diamond," Jason said.

"You're kidding, right?" Even Bill's optimism knew limits.

"Actually, no." Jason stretched as much as the cramped dayroom would allow. "It's an easier problem than nanite-grown hulls. I don't know anyone making microbots from diamond, but synthetic diamond *is* used for special-purpose computer chips.

"Carbon atoms are smaller than silicon, so we can shrink most everything in the 'bot. Expand it back to its current size, and there'll be some uncommitted surface area for photocells."

Jason warmed to the idea, his words tumbling out faster and faster. "Another thing. Diamond bearings will mean less friction in the micromotors. That'll reduce power use."

Bill emerged from the galley with the inevitable coffee bulb. "Can your friends Dirtside make diamond 'bots for us?"

The K-State labs had the necessary equipment; motivation was the question. His buddies knew only that he was away with Sherry—and when he went back, they expected to hear all the prurient details.

That was the least of his worries. He had more serious matters to lie about.

Jason thought about his tech-junkie friends back on Earth. He considered the practical challenges in making a diamond 'bot. The *fun*.

"Yeah," he said, smiling. "They'll do it. Count on it."

◆❖◆

A package for Jason arrived from Dirtside. An imperious summons followed close after.

The Syndicate heralded its messages, no matter how trivial, with high-priority tones. The ominous brass cadenza preceding *this* message had to signify a communication from a Syndicate bigwig. With a shiver, Jason opened the message.

Barbara Shaw, the station director, appeared. "I've scheduled ten minutes for Jason Grimaldi at 14:15 Station time." That was in twenty minutes, barely time to put on fresh clothes and reach her office. "Be prompt."

Shrewd eyes peered at Jason from the end-of-message still frame. He did not relish feeling that gaze in person. "Sherry, Bill … I can't explain that call. I've never met the woman."

To Jason's surprise, Sherry laughed. "You kicked up a stir Dirtside at the Syndicate regional office. Maybe Ms. Evil Eye means to resume the conversation you started with her campus recruiter."

He goggled. "You *knew*?"

"That you visited the local Syndicate office after I approached you? Yeah, we knew that. The receptionist is one of us."

"Then why am I here?"

Bill fielded that one. "Our agent listened in by intercom. You got into a nasty fight with the company man. You kept working with Sherry, so we decided you were on our side."

Jason looked away, red-faced. "Sherry gave me a lot to think about, but I had to make up my own mind about the Syndicate. You couldn't have known that. The scene could have been staged for your benefit."

Sherry squeezed his hand. "It could. I told Uncle Bill I didn't believe it had been. I trust you."

Jason had to know. "Why?"

She winked. "You talk in your sleep."

◆❖◆

The director took a call just as an aide escorted Jason in. She waved him to a seat, her attention on the call.

The butter-soft leather of the autochair molded itself to Jason. Its gentle embrace, not a mundane seatbelt, held him in place. Like everything in this luxurious office, the chair reeked of affluence.

*There's an interesting expression*, he thought. *Before Sherry, did I ever consider affluence capable of reeking?*

*Not important, Jason.* If this woman learned what he, Sherry, and Bill were up to, they would all be in the station's brig. *Could Sherry be right? Was this merely a continuation of his argument with the Syndicate recruiter? Why would Shaw bother?*

The office was oak-paneled, with plush oriental rugs over an Italian marble floor. Merely the chair that nestled him would cost his parents a year's salary. To lob such opulence up from the bottom of the gravity well? It was a brazen display of wealth. (*Display? Another interesting word choice.*) Jason could not fathom the mindset that valued such ostentation while a breathtaking view of Earth went unappreciated *behind* Shaw's desk.

At first glance, Barbara Shaw looked little older than Sherry. That had to be ReJuv shots; she could surely afford them. Her level, penetrating gaze spoke of many years in command. Despite microgee, Shaw's drifting, flowing hair somehow stayed coiffed. She probably spent more on her hair than he spent, Dirtside, on rent. He fought—unsuccessfully—the urge to squirm in the autochair. It buzzed softly, remolding itself to him.

Shaw finished her call. "Thank you for coming, Mr. Grimaldi." She tapped at her computer console. "Ah, there you are. Tough classes, good grades, significant dissertation project. And very resourceful."

Jason had not come expecting compliments. "Excuse me?"

"Picking a fight with a Dirtside recruiter to raise your visibility, then coming to a Syndicate Station. Ever since the system flagged your name on a passenger manifest, I've been expecting your call. I had begun to wonder if you'd gotten cold feet.

"Naturally, the system has followed you since *Tom Paine*'s arrival. Zero-gee sports"—the station's tourist annex was famous for them—"seemed a bit extravagant for a grad student." Shaw arched an eyebrow. "Although I imagine you and Ms. Nilsson have another zero-gee sport to divert you.

"When the cargo manifest for today's shuttle named you as receiver of a gemstone shipment, I understood the delay: careless packing." Shaw smiled broadly at him. "Gifts aren't necessary, of course, but I won't say no."

*Damn the guys anyway, declaring the diamond 'bots as gemstones.* No doubt they thought they were being clever.

How could the occupant of *this* office want a bribe from an entry-level applicant? "Uh … so Mr. Chrisp was negative about me?"

She laughed heartily. "You actually compared the Syndicate to the British East India Company?"

"Yes. Are you offended?"

"Flattered. Commercial control of the subcontinent pushed the state of the art in the 1700s."

*Like commercial control of a planet today?* The chair fought Jason as he leaned forward abruptly. "Then you admit it!"

"You have something against profits? That's *not* an attitude I look for in employees."

Jason said, "I'm normally all for profits. It's monopoly profits I question."

Shaw frowned. "What are you suggesting?"

"License your nanotechnology. Allow some healthy competition."

"Nanites give us a technological edge, Jason. Why give it away?"

It was wasted breath, but Jason felt compelled to try. "Licensing isn't giving it away. You could earn substantial fees without stifling all competition, *and* the solar system would get settled much faster."

She sniffed. "Licensing offers only a pittance. We'll make a lot more keeping the technology to ourselves."

"The technology should never have been exclusively yours. Just like the British East India Company, the Syndicate is a *political* monopoly. Oh, what you have is more subtle than a royal letter of patent, but it's a political creation none the less."

Sherry had spent years deconstructing a system of subsidiaries, joint ventures, shell companies, strategic investments, corporate alliances, political obligations, and research foundations. Jason summarized as best he could. "There's Syndicate funding behind every alarmist, lunatic fringe, anti-nanny group on Earth. Your 'edge' comes from hobbling nanite competitors more than from your own research."

He ignored her deepening glower. "The Smithson bill to stop nanotech work Dirtside will give you an insurmountable lead. Your ships will drive all the other companies, and all the independents, from space. And *then* won't we be surprised when prices rise—but not *quite* to the levels that would let another megacorp back into space."

Shaw leaned forward. "We're done, Mr. Grimaldi. With your attitude, I can't imagine why you bothered coming here."

Jason stood to leave. *And I hope*, he thought fervently, *that you never do.*

◆❖◆

The *Tom Paine*'s bridge felt hotter and more cramped than ever. Perspiration coated Jason's face, beads of sweat floating free from time to time, as he willed forward another tiny robot. They were working round the clock, certain that Director Shaw would invent a reason for canceling their berthing privileges.

Jason drove their tiny scout faster than he had ever before dared, telling himself better camouflage made that safe. Random defects induced on the diamond surface scattered light and suppressed reflections. The translucent material took the color of any surface on which the 'bot found itself.

And thanks to Bill's brainstorm, this scout had begun its cross-room scurry with its batteries fully recharged.

"Go, baby, go," Sherry urged.

The holo view crept along by another tiny step, and another, and another. The 'bot crawled along the wall to a corner. Then a second corner. Then down the third wall onto an instrumentation shelf. They cheered as the 'bot neared a testing port for the immense synthesis vat. The vat itself abutted the station, unseen beyond the bulkhead. Inside the vat the ninth Europa-class cruiser was taking shape.

Bill clasped Jason's shoulder. "A lot rides on this, son, for the planetful of people below and a few more offworld. They'll never know what happens here. But mostly I'm thinking of good folks who died in the Trojans, Sherry's parents and brothers among them.

"Sure, they came for the money, but that was only part of it. A small part. They believed humanity's future was in space. Not the Syndicate's future, but mankind's.

"We mustn't allow them to have died in vain."

Sherry hugged her uncle, leaving an arm draped across his shoulders as the 'bot took its final steps. The virtual bull's-eye lay perfectly centered in their display. In a corner of the image, a timer counted down.

The moment, when it came, would be too brief for human reflexes. Now everything depended on software.

*There.* A sampling port slid open. The automated quality-assurance probe stabbed into the aperture. Just as rapidly the 'bot thrust itself through the opening. They glimpsed dark turbulence before an eddy swept away the 'bot. The holo dissolved into static.

With one keystroke Jason switched their point of view. From the fly-on-the-wall perspective of the nearest relay, they watched the service port slide shut.

Jason squeezed the leathery hand that still gripped his shoulder. "Now, we wait."

◆❖◆

Macromolecules swam purposefully through stygian darkness, grazing on glucose dissolved in the murky fluid. The breaking of chemical bonds provided power; the carbon atoms liberated from the nutrients were retained as essential building material. One by one, their carbon receptors sated, the nanites switched to delivery/construction mode.

Molecular programs sensed and adapted to past progress. And while each nanobot worked alone, placing one carbon atom at a time, they numbered in the trillions. Together, they built with amazing speed.

A structure of marvelous complexity emerged.

Freed of their carbon cargo, programming sent the tiny machines back into the solution to gorge and build anew.

A process so massively parallel entailed many missteps. Only less-than-microscopic bits of the evolving construction ever came within range of any one constructor's chemical sensors. Decisions made from such limited data were often incorrect. A single atom bonded to the wrong place might miscue other constructors, and lead to more mistakes. Errors would propagate …

Quality assurance nanites prowled the border of the ever-expanding structure, comparing whatever they found to hull blueprints encoded in their molecular memory arrays. These fixers tagged and snipped out the anomalies that inevitably crept into the construction.

Scavenger nanites, like white blood cells, hunted foreign substances suspended in the solution. They engulfed stray contaminants and swarmed scraps tagged by the QA 'bots, before plunging sacrificially, clutching their prey, into a sticky entrapment filter.

A scavenger nanite, one of billions in the synthesis vat, encountered something unusual. It hunted the unexpected and its software anticipated many possibilities, but no programmed response suited *this* discovery. Diamond fragments were to be expected, but not with traces of phosphorus, boron, and aluminum. Creeping along the immense object, apparently millions of angstroms in length, the scavenger reached a decision. The object did not belong. The scavenger prepared to release a stream of chemical signal markers. It would need many helpers to engulf the diamond fragment.

It never got the chance.

A vortex formed in the fluid. Other nanobots, closer than the scavenger to the turbulence, were immediately swept away. The scavenger struggled, its flagella flailing, against the suction. Even at full power it could not resist. The nanite was sucked through a tunnel into a vast cavity. The tunnel disappeared, sealed by a diamond slab. Emergency programming took control and the scavenger darted in all directions without effect.

It took no solace from its chemically sensed company: a herd of constructors, a smaller set of fixers, and other scavengers. It did determine, as it reflexively swam and grazed, that enclosed with it were sufficient dissolved nutrients for a lengthy stay.

◆❖◆

*"Kashmir?"*

"Yes, Ms. Shaw," answered Keith Higgins, the director's flustered, balding deputy. "The cryptogram to Grimaldi's woman companion originated in Kashmir. Remember the spokesman for the Kashmir Liberation Front? Something Umar. He was at the independence talks in New Delhi. The tall, swarthy guy with the scruffy mustache. He's the new Kashmiri foreign minister."

Accessing the private message was illegal and contrary to Syndicate policy. Barbara Shaw shrugged. Self-defense was a higher policy.

She could picture Nayeem Umar, all right. So: The local malcontents were dealing with a Third World terrorist. Clearly, she should have taken more seriously the name *Tom Paine*, in honor of the infamous colonial radical.

"Time for our own 'Common Sense.' Let's see that recording, Higgins." She wondered how much computing time had gone into cracking the cryptogram.

Umar's message was terse. "*Salaam*, my friend. I'm pleased to say the cabinet has completed its debate. We accept your Tweedledee proposal. Succeed, and the sum we discussed will be invested in your enterprise. Good hunting. *Allahu akbar.*"

*Tweedledee? Why*, Shaw wondered, *not Tweedledum?* She could never tell them apart.

In a way, that was the point. Newly independent Kashmir must be planning to bootstrap its economy through the methods pioneered by Singapore, South Korea, China, and Vietnam.

Unauthorized copying.

◆❖◆

Within the body of the swimming gnatbot, on nearly frictionless diamond bearings, tiny gyroscopes spun. Despite currents and turbulence in the synthesis vat, the gyrocompass maintained range and bearing to the access hatch used by the quality-assurance probe.

Its sampling complete, the 'bot swam back to the QA portal to await the hourly opening.

◆❖◆

Despite maximum magnification and computer enhancement, the three conspirators could see no sign of the overdue scout.

"Move closer," Bill urged.

Jason resisted the temptation. Any time the relay moved, it

chanced losing its grip on the bulkhead—and with it, any hope of relinking with the missing 'bot.

They wanted a better view. They *needed* an operational comm link.

"Sorry," Jason said. "It stays."

They stared and wondered and worried. The QA sampling had happened right on schedule. Their scout should have emerged then from the vat, clutching the probe needle. It should have reestablished communications minutes ago. Where was it?

An emergency siren howled. Jason twitched; only a loosely fastened seatbelt kept him in his chair. He slapped the ACCEPT key. "*Tom Paine*."

The synthesis-chamber anteroom shrank into a corner of the holo. A balding man with a smirk took its place. His blazer bore a Syndicate crest. "Good day. Captain Nilsson, please."

Bill leaned in front of the camera. "Speaking."

"Keith Higgins, here, Manager of Station Operations. I'm afraid, Captain, I have some inconvenient news for you."

"You're awfully wordy for an emergency," Bill said.

Higgins shrugged. "Your comm is queuing routine calls. Rather than leave a message, I thought I'd get your attention. You might appreciate the lead time."

Bill sighed. "Get to the point, please."

Higgins's sneer broadened. "Beta port is overdue for maintenance. At this time, our other ports are either busy or reserved for incoming Syndicate ships. Bottom line, the *Tom Paine* must undock by midnight, station time."

Midnight was scarcely ten hours away.

◆❖◆

In a corner of the holo, flickering caught Jason's eye. He passed a scribbled note to Bill: Follow my lead. To the camera Jason said, "That's impossible, I'm afraid. Our main radio blew its power transformer." He looked expectantly at Bill.

The captain nodded. "Right, Higgins. The transmitter is a discontinued Sony-Boeing model. The station doesn't stock parts for it."

"We were unaware of this problem," Higgins said.

Jason glowered. "Why would you be? Are you monitoring our messages?"

Higgins's smirk faded a bit. "No, no, of course not."

Meaning almost certainly that the Syndicate had. *Uh-oh*, Jason thought. *What about the message from Kashmir? Had it been sufficiently secure?*

Jason tried to put that out of his mind. They could do nothing about it now. "A friend from O'Neill Two passes through next week. He's scrounged a secondhand unit to bring us. Your routine maintenance will have to wait."

"Unacceptable," Higgins snapped. "Have the part sent up on the next shuttle. Under the circumstances, we'll pay the difference. You have twenty-four hou …"

Long enough. Jason thumped the CANCEL key. The flickering picture-in-picture zoomed to refill the display. "Look, guys."

The grainy cross-room view gave way to a blurry close-up of the sampling equipment. The new scene wavered before flipping back to the fly-on-the-wall perspective. The images toggled jerkily.

"See that power reading? I had to act fast." Jason typed frantically. "It's almost drained." It: the overdue scout.

"Where's it been? What's going on?" Bill demanded.

Jason kept his eyes on the console. "I'm still working that out. The 'bot must have come out on schedule. The QA port has only reopened the once."

Jason paused the image on a dim close-up of the QA probe. "Look how dark the scout's vision is. The nutrient glop is thicker than anything we'd imagined."

"It won't be just the eyes," Sherry said. "The glop must be all *over* the 'bot. Is that why it didn't reconnect to the comm relay? Gunk on the lasers and photodetectors?"

Jason nodded. "That's my guess. Without a link, the 'bot would've gone into default mode. That means jigging about randomly, trying to get away from a presumed obstruction."

"Fat lot of good that will do," Bill contributed. "It's *wearing* the obstruction."

*Uh-oh*, Jason thought. "Coating the photocells, too. The 'bot's not recharging."

Sherry studied the gauge. "So the random motion is draining the battery—what little juice it has left after an hour swimming in syrup."

Jason's fingers raced over the keyboard. They did not have much time.

Bill kicked off toward the airlock. "I can't contribute anything here, and someone has to order a new widget. I'll downlink from someplace very public and make a great show of it." The inner hatch sighed shut behind him.

"Done." Jason slumped in his chair. "I've knocked together a program tweak and downloaded it to the relay. If we're undeservedly lucky, the intermittent link will stay up long enough to deliver the new subroutines."

"Why is the 'bot returning to life now?"

"Best guess, Sherry? Some of the goop is evaporating. Maybe it's jiggled some off. If we can't recharge the 'bot soon, 'why' won't matter. It's got a long walk ahead."

If the little diamond robot never reached a public area of the station for retrieval, everything would have been for naught.

Jason took Sherry's hand. "*If* the new program gets installed, we have a chance. It'll send the 'bot walking straight up the nearest bulkhead toward a ceiling fixture. Light is trickling into the photocells, just not enough to fully replace the power it's using. The closer, the better."

Sherry squeezed Jason's hand. "The closer, the better," she agreed.

◆❖◆

Infrared flashes, insistent but meaningless.

The gnatbot scout could not localize the source. What signal it received was oddly weak. It could not sense the nutrient residue over its photodetectors, dimming the IR laser signal.

The little machine kept transmitting in the best-guess direction. The feeble incoming signal implied a distant source, so it boosted power to its comm laser.

The draining of its battery accelerated.

◆❖◆

Side by side, silently, Jason and Sherry watched the downward crawl of the simulated power gauge. The simulated needle kissed the redline: five percent power remaining. The rate had slowed, but the image still flickered.

Nails pinched Jason's hand as the sampling port, glimpsed in flickers, began slowly to swing from sight. The view shifted by thirty degrees, then forty-five. It kept on turning. He did not breathe again until the glacial turn revealed the back wall. Then he checked the power reading. Then he cheered. After a moment's delay, Sherry joined him.

The scout 'bot had accepted the new programming. It was headed across the shelf toward the bulkhead for its long climb upward to the light.

## *Epilogue*

New hulls grew in three enormous floating microgee vats, soon to join the ten deep-space vessels docked to the Independent Miners orbital station. Cargo lighters, passenger shuttles, orbital utility craft, and spacesuited figures jetted everywhere. Tethered bales of nanite-grown diamond struts and panels for the station's continual expansion floated everywhere. Luna, austere and majestic, was two hundred klicks below, almost close enough to touch.

And the colonists below were close enough to quickly reach *here*. Away from Earth's bureaucracy meant away from the rule of law. No overt action had ever been taken against this station, and perhaps the proximity of the helium-3 mines—and hundreds of independent miners—was only a coincidence.

And perhaps not.

But this was not the day for negative thoughts. Jason stood to go meet his visitor, the new CEO of the Syndicate, in the station's main lounge. He had not quite reached the office door when Barbara Shaw burst in, with Jason's apologetic-looking deputy trailing after.

The brisk *zzzp-zzzp* of Velcro footwear turned Shaw's grand entrance almost comical. She ignored Jason's outstretched hand. "Grimaldi, you cannot imagine how I've anticipated this day."

"I'm sure you'll tell me. Why don't we sit down first?"

She remained standing. "Nice office. Think you can still afford it?"

From what Jason remembered, this cabin could not hold her old desk. His token nod toward personalizing the office involved a couple liters of paint and some holo art. "Can I?"

"My accountants predict that you'll squeak by. That proved to me that my lawyers were not sufficiently … motivated. I fired them."

Sighing, Jason grabbed two empty drink bulbs from a cabinet. "Coffee?"

The hospitality made her blink. "Don't you realize I did my best to ruin you and your precious Independent Miners? You stole Syndicate nanites—somehow—and used them to jumpstart your own corporate empire. It's taken me five years to establish your piracy, but now I've stopped you in your tracks."

He filled both bulbs and handed her one. "Look, Barb"—she bristled at the familiarity—"you haven't *established* anything. The Independent Miners settled out of court. True, the Syndicate will get a lump-sum payment now, and royalties on our future ships, but we've admitted no wrongdoing."

"And the ships that are indistinguishable from half my fleet?"

Jason gestured at his window, its polarization set high to soften the lunar glare. "Look two docks over. That's the *Madrid*, the ship you rode out here. It mates easily with our docking ports and our fueling rigs, and we can service pretty much any part of it. And although it probably galls you, our ships sometimes dock at Syndicate stations. They can refit and repair with parts from your depots. Standardizing on size and shape is simply practical."

In her right temple, a blood vessel throbbed. "And there was no significance to the name of your first diamond ship? Do *not* take me for a fool!"

"Ah, the *Growing Paine*. I'll tell Sherry you appreciated her little joke."

"The monetary settlement is no joke," Shaw snapped. "I came to watch you transfer the funds, and I want to see that *now*. We're meeting here, Grimaldi, rather than in my office, for one reason. So you can't pretend that paying up doesn't hurt."

Shaw didn't get it; maybe she never would.

"Look outside first, Barbara. Tell me what you see."

"A space station, some construction, modern ships—if all, like this office, rather Spartan. Nothing I didn't see as a station exec. What's your point, Grimaldi?"

Twice as many ships were here today as Jason had seen at Syndicate Station Three, but he chose not to quibble. "We're opening the solar system faster than the Syndicate ever would have—certainly faster than it once chose to. These ships, and more like them, bring us all—Earth, the Luna settlements, the O'Neill colonies, Mars base—far cheaper resources than five years ago. And they're giving *you* a bit of honest competition." (Shaw snorted at "honest.") "From its last annual statement, I doubt the Syndicate has suffered much from vastly expanded markets."

Jason smiled. "And, at least after the fact, every one of our new ships was built under license from the Syndicate."

"Under license?" she scoffed. "You once sat in my office and dared me to license our technology. You planned this even then, down to settling out of court. It's very clear now. Admit it."

Jason turned to admire the nearest synthesis vat. "Clear? A new cruiser is growing inside, nanoseeded the day your lawyers and ours reached agreement.

"We call it the *Window Paine*."

# *The Next Small Thing*

Throughout human existence, humans have made things—whether spears or spaceships—by manipulating matter in bulk. We're not as different as we might care to believe from cavemen chipping away at chunks of flint.

How we make things is about to change, as engineers begin to manipulate individual atoms. Think of the new technology as embracing quantum weirdness. The transformation to building things with atomic precision will be as revolutionary—in my opinion, in any event—as the Agricultural Revolution, the Industrial Revolution, or the Information Revolution.

And revolutions are, invariably, Interesting Times.[1]

Read on for a peek into the coming Nanotechnology Revolution.[2]

---

1. Interesting times, even as they are turning our lives upside down, are the novelist's friend. The research that appears in this section began in data-gathering for *Small Miracles*, a novel of medical nanotech.

2. "Small Business" could as easily have wound up in this part of the collection as in the "Out of the Cradle" section. Consider the story surrounded.

## *Follow the Nanobrick Road*

### (First published in September 2008)

K. Eric Drexler's 1986 science popularization *Engines of Creation: The Coming Era of Nanotechnology* was a real eye-opener for me. [1]

By two decades later, nanotech had become big business. And too important to be left to business: The federal government (and many governments around the world) invests heavily in related R & D. The National Nanotechnology Initiative has spent $6.5B over five years, with spending continuing to ramp up. A National Science Foundation expert foresees a trillion dollars in annual worldwide production of nanotech-related goods by 2015.[2]

### *What is nanotech?*

Nanotechnology is most simply defined as engineering or manufacturing performed at the scale of nanometers. A nanometer, abbreviated nm, equals one billionth, $10^{-9}$, of a meter. For those who prefer classical standards of reference, human hairs vary in diameter from about 10,000 to about 100,000 nm.

"Small stuff" is insufficient to explain the allure of nanotech, and the billions in public investment. While it's easy to find products

1. The concept of nanotech predates *Engines* by a bit. Experts cite a 1959 address by physicist Richard Feynman to the American Physical Society. (See "There's Plenty of Room at the Bottom," a copy of which is posted at http://www.zyvex.com/nanotech/feynman.html.)

2. The study cited in the original (2008) link is no longer available. Here's a 2010 study that offers the same forecast: http://wtec.org/nano2/docs/ChaptersPdf/Ch0_2_LTV_10-1121.pdf

labeled as nanotech, in categories from:

- sunblock (with active ingredients in the form of 30-nm particles of zinc oxide), to
- clothing (whose fibers have been nanocoated for stain resistance or antistatic properties), to
- golf clubs (made lighter and more flexible by embedding carbon nanotubes in the composite material of the shaft),

what really excites people is the potential for far more sophisticated applications.

Individual atoms are tenths of a nanometer in size. (Not that atoms have precise boundaries, due to pesky quantum-mechanical considerations. Quoted sizes for atoms derive from interatomic separations within molecules.) The allure of nanotech is *atomically precise* engineering, and not merely of static materials. The title of Drexler's book speaks of engines, not golf clubs. Now imagine what might be done with truly tiny engines built with atomic precision ....

### *The road to nanotech*

October 9-10, 2007, I attended a conference organized by the Foresight Institute (cofounded by Drexler in 1986), Battelle Labs (the not-for-profit R & D entity that administers several national labs), and the Society of Manufacturing Engineers. This article draws heavily, though not exclusively, on the conference to illustrate the progress and promise in nanotech.

An ambitious purpose was reflected in the conference title, "Productive Nanosystems: Launching the Technology Roadmap." We'll return to that curious qualifier: productive. As for "roadmap," the antecedent is to one of the most successful case histories of industrial policy, the International Technology Roadmap for Semiconductors (ITRS).[3]

Manufacturers and customers alike grew reliant on predictable

3. http://www.itrs.net/

advances in integrated-circuit technology, and the promise that chips will steadily grow more feature-packed (and hence faster, smaller, and cheaper). We call that forecast Moore's Law, but it's hardly a law of nature.[4] Maintaining that rate of progress takes a concerted, ongoing international collaboration among corporations, academia, and governments. Semiconductor foundries take years and billions of dollars to build—a misstep can doom a company. It's not surprising that the ITRS is actively supported and regularly updated.

Nanotech is a young field, far less mature than semiconductor electronics at the time of ITRS's conception. The nanotech roadmap is a daunting—and potentially invaluable—endeavor. Because nanotech *is* so young, to choose a path forward is challenging indeed.

### *One (tiny) step at a time*

Our journey begins with *nanomaterials*. As we've seen, some nanomaterials have already found their way into products. Not all nanomaterials require atomically precise manufacturing.

And not all applications are as inconsequential as golf clubs. Nanometer-thick films enable the giant magnetoresistance (GMR) technology that crams ever more gigabytes into disk drives. ("Giant" refers to the significant decrease in electrical resistance that can be induced in nanometers-thin alternating ferromagnetic and nonmagnetic metallic films. The GMR label dates back at least to a 1988 paper in *Physical Review Letters*.) Denser storage is rather handy for laptops, to cite just one benefit. Peter Gruenberg and Albert Fert

4. Gordon Moore remarked in 1965 that the number of transistors on a chip had roughly doubled every two years. He boldly predicted the trend would continue.

In 1965, cramming sixty transistors onto a chip was a research project. In 2006, Intel (an apt example, since Moore cofounded Intel) introduced its Core Duo microprocessor line with *291 million* transistors on a chip. 2006 – 1965 = 41 years, in which Moore's Law would suggest 20.5 doublings. 60 X 220.5 ~88 million. Not bad, considering that virtually every aspect of transistor manufacture has changed, some repeatedly, in forty-one years.

Chips didn't get appreciably larger, meaning that individual transistors have gotten a lot smaller over time.

received the 2007 physics Nobel for their work with GMR.

Next, consider the single-wall carbon nanotube: a sheet of graphite (a form of carbon) just one atom thick, rolled into a cylinder only one nanometer in diameter. A carbon nanotube can, in theory, extend to any length. A single-wall carbon nanotube is stronger and lighter than steel, and an excellent conductor of electrical current. (Not all carbon nanotubes take the single-wall form. In multi-wall forms, carbon nanotubes can be concentric cylinders and rolled-up scrolls.)

Despite their nanoscale aspect, nanotubes are manufactured by such brute-force methods as laser ablation and arc discharge. We can hope to weave nanotubes into a superstrong cable for a space elevator, but record-breaking nanotubes are only a few centimeters long. No space elevator for us *this* week ....

That's our starting point. Where are we going?

The roadmap leads us first to *nanocomponents*. These are building blocks like motors, pumps, bearings, gears, and sensors. Nature builds nanocomponents; surely we can, too. (A bit on how comes later in this article.) Over the next decade, it is expected researchers will build an extensive library of nanoscale components.

Interesting things can be done with just nanocomponents. David Leigh, of the University of Edinburgh, introduced a surface coated with the simplest of nanoscale motors: bistable molecules. In one position, the molecules are hydrophobic (water-repelling), and in the other, hydrophilic (water-attracting). Ultraviolet light flipped the molecules between states. Leigh showed a video of a one-*milli*meter ($10^{-3}$ meter: far larger than nanoscale) water droplet on this surface. UV illumination impelled the droplet across the surface—even up a significant slope.

The next stage (five to fifteen years from now?) is *functional nanosystems*, combinations of nanocomponents. These systems will transform materials, information, or energy. They just won't transform *much*. Typical production quantities are predicted to be in the milligram range.

Next up (ten to twenty-five years out): *atomically precise produc-*

*tive nanosystems.* APPNs will introduce a new era in nanotech and manufacturing. APPNs will *produce* (hence the conference name) other objects with atomic precision, and do so under programmable control. The programmable control need not be onboard the APPN. By reason of safety or public relations or efficiency, it may be best if the programmable control not reside in the APPN.

(Of course we'll talk about nanobot replicators. In a few pages, I promise.)

Squinting into the metaphorical distance (fifteen to thirty-plus years out) we come to *scaled, atomically precise, productive nanosystem array systems.* With these we will finally build large objects, and bulk materials, with atomic precision. Carbon nanotubes are stronger than steel because they are defect-free, not because of nanoscale. Atomically precise manufacturing will let us build *any* material without defect—and then *steel* will be much stronger. Many materials can, in theory, be made two orders of magnitude stronger by suppressing defects.

| **Overview: Nanotechnology Roadmap** | |
|---|---|
| Nanomaterials | 2007 |
| Nanocomponents | 2 – 10 years |
| Functional nanosystems | 5 – 15 years |
| Atomically precise productive nanosystems | 10 – 25 years |
| Scaled APPN array systems | 15 – 30+ years |

### *Why so vague?*

The preceding section may seem less of a roadmap than a rough sketch of the territory ahead. Fair enough—and with good reason. Much of the conference dealt with alternative techniques for the first step forward: making nanocomponents.

Several methods offer promise. At the most abstract level, these boil down to biochemical and mechanical approaches—popularly, wet and dry.

Living cells teem with sophisticated molecular machinery. DNA encodes the directions to build complex and quite large structures (like us). Biology proves by example that nanotech is possible, and biologists and biochemists presented a fair number of papers at the conference.

Alas, nature can be a bit of a sluggard,[5] and many chemical processes yield unwanted byproducts. There's appeal to purposeful control over every step of building things.

Wet vs. dry was *not* an open issue at the conference, and Drexler cautioned in his address that it's too early to pick an approach. He stressed, in fact, that the different techniques may complement each other.

### *That does not compute*

It is challenging to realistically model even simple molecules.

Our senses delude us that objects have properties that can—at least in principle—be measured with absolute precision. Attributes of objects, such as their size, position, and velocity, seem continuous and infinitely divisible.

At sufficiently small scale, our intuition ceases to be valid. Electrons and nuclei are governed by quantum effects that manifest only at tiny scales. For example, the Heisenberg Uncertainty Principle tells us that location and momentum cannot both be measured to infinite precision. The more one knows about location, the less can be known—even in principle—about momentum. And vice versa.

Schrödinger's equation describes the theoretical behavior of particles like electrons and atomic nuclei. The descriptions are probabilistic rather than determinate. Except in very artificial circumstances (like a lone electron in an infinitely deep potential well), the equation lacks exact solutions. Quantum chemists model the atomic behavior of atoms within molecules using computation-intensive approximations.[6]

5. The Earth formed about 4.5 billion years ago. About one billion years later, living cells had established themselves. Nature needed about four billion years to get to multicellular life.

6. For an introduction to quantum chemistry and the role of computation, (warning: it's heavy-duty math) see http://web.archive.org/web/20050308195257/http://cmm.info.nih.gov/modeling/guide_documents/quantum_mechanics_document.html.

Hence: The electrical forces at work among atoms (nuclei attract electrons; nuclei repel nuclei; electrons repel electrons) can be characterized only probabilistically and inexactly. The best approximations are computation-intensive, manageable for only a few atoms. Quickly enough, modelers fall back on nineteenth-century physics to do much simpler—and less precise, of course—approximations.

How many atoms might a nanoproduct contain, and hence might we desire to model? Even for very small items, a lot. A buckyball is a hollow carbon sphere about 1 nm in diameter—comprised of 60 carbon atoms. A cube (of most anything solid) just 5 nm on a side contains tens of thousands of atoms.

The need to work with approximation upon approximation is a significant hurdle to our understanding. Nano*tech* is out ahead of nano*science*.

### Ramping up

The roadmap envisions eventually building macroscale objects with nanotech. Defect-free material will enable new structural applications—perhaps, finally, that space elevator. Defect-free conductors can make our power-distribution systems more efficient and reliable.

However …

A few grams of matter contain roughly a trillion trillion atoms, and we can't precisely model the behavior of two.

Computation is not the only approach, of course. David Geohegen of Oak Ridge National Labs spoke of the need for better manufacturing-level understanding of nanomaterials. Focusing on single wall carbon nanotubes, Geohegen summarized the gaps in our knowledge of how the tubes grow and what circumstances introduce flaws. He advocated for standards (e.g., to characterize defects and purity) in nanotube production.

### Fun with atoms

Manipulation of individual atoms was first demonstrated in 1990 by Donald Eigler and Erhard Schweizer. Their employer cannot help but spring to mind. To demonstrate their ability to move single atoms using a scanning tunneling microscope (STM), Eigler and Schweizer released an image of thirty-five xenon atoms set onto a nickel surface, arranged into letters: IBM.

Atoms are *small* (xenon atoms being on the heavy end, each weighing about 1.31 X $10^{-20}$ grams). Building atomically precise products one atom at a time will take … *lots* of time.

Integrated circuits with very fine details are produced through a technique called photolithography. Light is projected through extremely detailed masks to expose a chemical film, called a photoresist, previously deposited on a silicon surface. Light-exposed photoresist hardens to protect the surface; unexposed regions are then chemically washed away. The resistless areas can then be further treated (e.g., to deposit metallic interconnects) without affecting still covered regions.

One method much discussed at the conference was *atomic* lithography. Rather than paint design patterns with light, atomic lithography uses the very fine probe tip of an STM to dislodge individual hydrogen atoms from the silicon surface. Michelle Simmons of the University of New South Wales netted in from Australia to report having built interconnects just 2.5 nm in length. In contrast, a state-of-the-art semiconductor plant makes chips with features no smaller than 45 nm.

Moore's law suggests industry will be manufacturing chips with atomic-scale features by 2020. Simmons's team may be blazing the trail. They have used atomic lithography and STMs to precisely implant individual phosphorus atoms into crystalline silicon. The up/down spin states of individual atoms (rather than, today, bunches of electrons stored in tiny capacitors) may eventually encode digital bits.

The next step is to marry atomic lithography with parallelism. Thomas Theis of IBM discussed a prototype 4000-bit read/write head for very high density (terabit/inch$^2$) disk drives. That's getting close to thousands of individually addressable, program-controlled probe tips.

Nature is nanotechnologists' proof by example that matter can be organized en masse by atomically precise methods, but of course we don't find STMs in nature. For another path to nanotech, we turn next to some of the biologists' presentations.

### *Nature's way*

Many proteins self-assemble into complex structures, including—a *very* handy type of device—motors. Consider the whiplike structure called a flagellum that propels many bacteria. Amazingly, forty proteins self-assemble into a structure of bearings, rotor, drive shaft and whip. ATP molecules (the tiny energy carriers within cells) break hydrogen bonds in the tiny engine to drive a rotary motion that flails the whip end.

Let's look more closely at self-assembly.

Diffusion brings components into proximity—within the tiny volume of an E. coli bacterium, it's been calculated every two organic molecules meet about once per second—and then non-bonding electrostatic attraction may assemble something larger. Randomly colliding components won't always orient properly to connect, but soon enough the flagellum *does* come together.

As another example, kinesins are motor proteins that pull themselves—and cargo—along intracellular tracks called microtubules. The kinesin molecules exploit thermal motion within the cell to take 200-300 "steps" per minute. This is one of nature's neat tricks: using random thermal motion to drive motion in one direction. Symmetry-breaking mechanisms—effectively like a ratchet and pawl—favor selected impacts, giving rise to a directional bias.

Biological motors can be quite powerful. Keith Firman of the University of Portsmouth (UK) showed a video of a 20-nm device, driven by a light-powered proton pump, crossing 70 microns. That's 3,500 times its own length.

Naturally occurring biomachines may serve as nanocomponents, but so might engineered biomachines. Walking and twisting motors built from DNA strands have been simulated.

Perhaps mechanical and biological mechanisms will be combined. For example, carbon nanotubes might serve as tracks for kinesins.

It's unlikely that naturally occurring proteins will fit our every need. Then what?

Custom-designing a protein is a hard problem. Proteins consist of multiple peptides—each peptide, in turn, comprised of multiple amino acids, themselves complex. A protein's function depends less on its composition—so many carbons, oxygens, hydrogens, etc.—than on the precise structure and shape of the molecule. The protein folds as it assembles, subject to the complex interplay of electrostatic interactions (and quantum uncertainties) between its many parts. On an IBM Blue Gene L, among the fastest supercomputers in the world, to model the folding of one protein can consume weeks of computing time.

Why? Vibrations between neighboring atoms take on the order of $10^{-15}$ seconds, yet proteins need on the order of milliseconds (order of $10^{-3}$ seconds) to fold. There's a lot of random bumping about before the protein settles into a stable, energy-minimized shape. Any simulation must deal with trillions of time steps.

(December 2011 update: With advances in computing, protein modeling has become much more manageable. In October 2010, a special-purpose supercomputer modeled a protein folding within a millisecond.[7] Meanwhile, tasks that once daunted supercomputers are increasingly tackled by armies of borrowed PCs. And so, Folding@Home has succeeded, in at least one instance, at modeling a protein assembly within two milliseconds.[8])

Rather than design proteins from scratch we may modify existing ones. Replacing a binding site(s) modifies the protein's bonding behavior without altering its overall shape—hence, leaving intact the protein's structural properties.

Peptides often connect with single bonds, and single bonds can (and do) twist. Twisting is much of why peptide chains take so long to settle into a stable configuration, and why protein formation is hard to model. Christian Schafmeister of Temple University has

---

7. http://www.nature.com/news/2010/101014/full/news.2010.541.html

8. http://folding.typepad.com/news/2010/01/major-new-result-from-foldinghome-simulation-of-the-millisecond-timescale.html (The embedded video is well worth watching.)

built artificial peptides, called bis-peptides, that form double bonds. Bis-proteins assembled from bis-peptides *don't* twist.[9] Bis-proteins assemble more quickly than natural proteins. The behaviors of bis-proteins are more readily simulated than those of natural proteins, whose shapes remain mostly unknown.

How might protein-based components, whatever their source, be mass-produced? Schafmeister discussed an electrochemical synthesis vat, in which a computer mediates every step in the self-assembly of complex structures. The PC sends electrical activation signals to an array of catalysts. One catalyst enables each assembly step. Other catalysts drive disassembly of unwanted byproducts and excess intermediate components. Of course first we need to design and synthesize the catalysts ....

So: The good news is that research offers several promising approaches. It will take a while to sort out which methods work best for which tasks. Until nanocomponent technologies are more settled, we're unlikely to see with much clarity to the later stops along the road.

### Hey! What happened to replicators?

Chemistry (gazillions of jostling molecules) and biology (gazillions of reproducing cells) operate with massive parallelism. *Engines of Creation* gave us another vision of parallelism: the universal assembler.

Assemblers were nanoscale machines that could build—well, anything. Anything, that is, for which: raw materials were at hand, energy was available, and directions were onboard.

Most products, like that sports car I covet, would require that assemblers cooperate. *Lots* of assemblers. And how would we get all those assemblers? From assemblers assembling more assemblers, assembling more assemblers ... until there were enough for the job.

Fiction has a sexier name for self-assembling robots (both nanoscale and larger): replicators. With sunlight for energy and the

---

9. "Molecular Lego," *Scientific American*, February 2007.

entire world for raw materials, it's easy to construct entertainment—or alarmist scenarios—in which replicators just keep replicating. Before you know it, there's nothing *but* replicators. And in these scenarios, what of humans?

Worst case, we're raw materials.

You will have noticed: Nothing even vaguely assembler-like appears in the roadmap. Avoidance of negative press may have influenced the choice of direction, but there is a more basic reason. To reap the benefits of nanotechnology does not require anything as complex as an all-purpose, self-replicating, autonomous assembler.

Two decades ago, Drexler envisioned teams of assemblers at work within synthesis vats and tabletop factories—not in the wild. Regardless, in the wild and living off the land is how much of the public (and some fiction) perceives assemblers. Asked at the recent conference about replicators, Drexler characterized that early description of nanotech as a biological analogy and proof by example. He said he considers *Engines of Creation* obsolete.

Forget living in the wild—useful nanofactories won't even need vats of mobile assemblers. Chris Phoenix, director of the Center for Responsible Nanotechnology, laid out a case against mobile assemblers in a 2004 interview.[10] Stationary assemblers—think: tiny robotic arms—are simpler and more productive than free-floating assemblers. It's simpler to deliver energy and raw materials to fixed assemblers. Fixed assemblers are easier to physically orient to the task than are free floaters. Prepositioned assemblers are easily coordinated by a central computer through permanent comm lines (perhaps carbon-nanotube wires). It is much harder to maintain communications with mobile assemblers, which would not trail cables lest they tangle.

A fixed 'bot that's taken out of its factory won't harm anything.

Basically, replicators are a hard problem, more complex than alternate nanoscale manufacturing approaches, scary to much of the public, and off the community's roadmap.

10. http://www.crnano.org/interview.phoenix.htm#Part%202

But are mobile, independent, self-replicating assemblers *possible*? In theory, sure.

Fear of gengineered superbugs predates most qualms about nanobots replicating. Bioengineers allayed many concerns through a program of community self-governance informed by dialogue with ethicists. Ethical guidelines, peer reviews, and funding reviews have kept genetic engineering safe and beneficial.

Nanotech researchers are taking the same approach. The Foresight Institute and the Institute for Molecular Manufacturing have proposed ethical guidelines for the development of nanotech.[11] In a nod to precedent, these guidelines, presently at version 6 (and still so in 2011) note: "The NIH Guidelines on Recombinant DNA technology are an example of self-regulation taken by the biotechnology community over 30 years ago." Among other topics, the guidelines address alternatives to autonomous, self-replicating assemblers. Should autonomous, self-replicating assemblers be developed anyway, the guidelines argue for layered security measures to counter accidental or purposeful release. A few examples of safeguards:

Replicators can be designed dependent—say, for their fuel—on chemicals unavailable in natural environments. This is like designing a genetically modified bacterium to starve outside the lab for lack of a rare trace element.

Material-scavenging functions can be limited to specific materials not found in nature, or not found in living cells.

Onboard programming can defend against software errors—analogous to mutations in a bacterium—such as with error-correcting codes.

But could someone *intentionally* design replicators to survive in the wild? Again, sure—given *lots* of money, because replicators are a tough problem. If it's any consolation, there are probably much easier ways to do us harm.

Even replicating nanobots designed with a malicious intent may not prove fatal. We manage to share Earth with a lot of ever-mutating, mindlessly replicating bacteria.

---

11. http://www.foresight.org/guidelines/current.html.

***Are we out of the woods?***

Let us suppose we have sidestepped, at least for the near future, the gray-goo scenario of replicators run amok. Does that mean nanotechnology is safe?

To be determined.

Nanotech's big uncertainty is a matter for toxicologists. Materials often exhibit different properties, chemically and biologically, at the nanoscale than in bulk—and yet some nanomaterials have been introduced to factories and mass-market products without performing new health tests. To infer a substance's safety from bulk-level testing while touting the differences manifest only at nanoscale … does that strike anyone else as a gamble?

Toxicologists are starting to consider effects unique to the nanoscale. For example, it's been noted that materials long deemed safe could never before—until formed into nanoparticles—reach the most sensitive parts of the lungs.[12] The Environmental Protection Agency recently asserted the authority to regulate—as pesticides—the nanoscale silver particles some new clothes washers release to kill bacteria.[13]

During a conference lunch, an insurance-industry representative brought up asbestos. Asbestos is a useful material whose health implications (asbestosis and mesothelioma) went unrecognized for decades. Related health science was behind the curve. Insurers were taken entirely by surprise. Class-action suits bred like, well, poorly designed replicators. Dozens of companies somehow linked to asbestos went bankrupt, some perhaps deservedly, but others perhaps the victims of junk science and improper litigation.[14]

---

12. "Nano Hazards: Exposure to minute particles harms lungs, circulatory system" reports on a related discussion at a 2005 Society of Toxicology meeting. http://www.sciencenews.org/articles/20050319/fob1.asp

13. http://bits.blogs.nytimes.com/2007/09/26/samsungs-washers-regulated-as-a-pesticide/

14. http://web.archive.org/web/20080516205439/http://www.asbestoscrisis.com/legal.asp

Let's hope nanopollutants aren't this century's version of the asbestos surprise.

(Let's keep hoping. In 2011, as researchers experiment with nanotech-enabled food wrappers to keep oxygen out and the food within fresh, "The federal government—the U.S. Food and Drug Administration, and to some extent the Environmental Protection Agency—has essentially no specific policy on nanomaterials and food, either in edibles or in packaging."[15])

### *On the sunny side of the nanotech street*

I'm not an every-silver-lining-has-a-cloud sort of guy. Let's turn our attention to plucking some low-hanging fruit along the nanotech road.

The roadmap focuses on energy and healthcare as opportunities for early benefits. We'll start with a few energy-related opportunities. Fuel cells that use catalyst nanoparticles to boost efficiency. Photovoltaic cells embedded with quantum dots to extract energy from wavelengths too short to be tapped by conventional solar cells. Cables and wires with fewer molecular-level defects—less electrical resistance—to distribute power with lower losses. Ultracapacitors, using the very high capacitance of nanoporous electrolytes, to replace batteries in many applications.

Some short-term healthcare opportunities? Nanoshells for improved control over drug delivery. Nanoparticle contrast agents for more sensitive MRI imaging. Engineered protein sensors for earlier detection of tumors. Engineered proteins that attack tumor cells. Engineered protein sensors that detect and report—as by fluorescing or changing colors in the presence of—biohazards. (Cellular-repair nanobots, should they come about, will be a longer-term undertaking.)

Nanotech is receiving a lot of public funds. It may face a headwind of public skepticism. An emphasis on applications with obvious public benefit and visibility is sound strategy.

---

15. http://www.newhavenindependent.org/index.php/archives/entry/small_bites_bigger_questions/id_41914

### *Stronger and smarter*

Let's be a bit more speculative. (And while we're at it, we'll change metaphors. We've left the roadmap conference, to gaze instead into our nanotech-enhanced crystal ball.

Fair warning: This isn't a comprehensive forecast—nanotech will change *everything*. That said, how might nanotech influence science fiction in the next few years, and our lives over the next few decades?

Many payoffs will come from structural nanotech: vastly improved materials.

Today's metals are rife with voids, cracks, and impurities. Metals without defects will be stronger by about two orders of magnitude. Imagine how much weight can be taken out of cars and trucks (and rockets) with no reduction in safety—and with major gains in fuel efficiency.

Big weight reductions will also come from substituting other materials for metals. Boeing, in its 787 Dreamliner, replaced metal panels with fiber-reinforced carbon composites. (The carbon fibers used are a few microns across, not technically nanotech.)

So there you have the nanotech-enabled auto of the future: electrical, powered by ultracapacitors or fuel cells, and weighing perhaps a few hundred pounds.

An SF standby is the electromagnetic cannon. These weapons come in two variations, railguns and coilguns. Their peaceful counterpart is the electromagnetic mass driver. All three use EM pulses, rather than chemical reactions, to launch projectiles. It's a very elegant notion, compelling demonstrated decades ago.[16] So why don't we have them?

Because the harder one hurls the projectile, the fiercer the electromagnetic forces exerted on the launcher itself. Today, those forces tear the launcher apart. But defect-free copper is approximately eighty times stronger than the natural stuff. Maybe we'll yet see rapid-fire railguns like those on *Stargate Atlantis* and *Battlestar Galactica*.

---

16. "For Love of a Gun," *IEEE Spectrum*, July 2007 (online at http://www.spectrum.ieee.org/jul07/5296)

Stronger is good, but we can also make material smarter. Materials coated or embedded with nanosensors can announce defects. Such sensors may be as simple as molecules that, when under mechanical stress, generate a voltage by piezoelectric effect or change their color. Nanoparticles may migrate within smart material to seal tiny cracks before they can grow.

David Leigh (recall ultraviolet light chasing a water droplet uphill) commented, "Nature uses controlled molecular motion for everything." How about using nanotech to give materials active surfaces? Paint hulls with bistable (hydrophobic/hydrophilic) molecules, and the ocean itself will push/pull ships. Nothing says only UV can flip such surfaces. To steer or change speed, merely tune control fields (electric or magnetic) or heaters in different parts of the hull.

### *Moore's Law marches on*

We've noted that a state-of-the-art semiconductor foundry makes chips with 45-nm features. We've discussed some of the experimentation leading the way to yet smaller features.

It's not a big leap to predict that our computers will use denser, higher-capacity random-access memories. But RAM in the future may not be wholly electronic. Fraser Stoddard of UCLA has demonstrated electrical readout of biomechanical RAM—an array of bistable molecular switches—with a density of $10^{11}$ bits/cm$^2$. You read that correctly: 100 gigabits per square centimeter.

Or, as suggested by the previously cited work at University of South Wales, memory may reside in the spin states of single atoms implanted into a semiconductor lattice.

Processor chips will continue to get smaller and faster, and again not necessarily by familiar means. No more futuristic than nanotech, but tiptoeing into quite different (and also potentially revolutionary) research ... quantum computers manipulate quantum bits. A memory bit in a conventional computer takes the value zero or one; a qubit may be zero or one—or both simultaneously. In quantum-mechanical terms, that "don't know" situation is called a superposition of states.

Ten qubits, to pick an arbitrary number, all of whose states are in superposition, simultaneously encompass 1024 (2 to the 10th power) possibilities. Thirty qubits can simultaneously encompass more than 4 billion (2 to the 30th power) possibilities.

It would take an entire article to properly introduce quantum computing[17] or why the spin states of single atoms implanted into a semiconductor lattice are candidates for implementing qubits for practical quantum computers.[18] It's worth noting, however, that the massive parallelism of fully realized quantum computing is well-suited to the computational challenges of further developing nanotech and to cracking the encryption schemes that underlie today's secure electronic communications. And that quantum cryptography, another subject too complex to tackle here, may make comm links secure again.

### *To sum up*

Nanotechnology is no longer coming—it's here.

Thoughtful researchers and practitioners have attempted to predict how nanotech will—and how it should—evolve. They see several paths forward and new applications soon to come within our grasp.

Atomically precise control of matter is a very young technology. It is, almost certainly, a transformative technology. The industrial age, the computer age, and next up—the nanotech age …

The one forecast I make with confidence is this: We will be surprised—often pleasantly, sometimes not—with the changes nanotech will bring in the coming decades.

### *December 2011 addendum*

Four years have elapsed since the roadmap conference. It's fair to wonder about progress over that time.

---

17. "Dot to Dot Design," *IEEE Spectrum*, September 2007 (online at http://www.spectrum.ieee.org/sep07/5489)

18. "The Diamond Age of Spintronics," *Scientific American*, October 2007.

How better to revisit a roadmap metaphor than with a few car-centric updates?

Sampling only quite recent reports, we'll start at the nanomaterial level. Consider, for example, a nanotech-enhanced electrolyte for batteries. Because the new electrolyte is fluid, rather than recharging your car battery you might drain the old electrolyte and refill the battery with previously recharged electrolyte. (In this vision, electrolyte swaps become the new function of the world's gas stations, repurposed.) Still, despite combining a fill-'er-up mode with energy densities comparable to the best lithium-ion batteries, this is a complex technology; automobiles may not be the initial application. Perhaps batteries enhanced with nanotech will first help solve the intermittency problem of solar and wind power, beefing up the national grid for plug-in cars.[19]

On to nanocomponents.

Researchers at Rice University have built a nanocar a mere four nanometers long. The car isn't self-powered—the probe tip of a scanning tunneling microscope powers each half-turn of every wheel—because this is a proof of concept. Still, it's good to know that a vehicle with such *very* tiny axles and wheels can roll rather than skitter about in random thermal motion.[20]

What about functional nanosystems, combinations of components? The roadmap foresaw progress five to fifteen years out: so, at the earliest, 2012. Alas, I can't put my finger on a good road-related example. Instead, consider a bacterium-targeting nanotech knife. IBM researchers have demonstrated nanotech devices that selectively bind to and then pierce bacterial membranes. Prospective targets include drug-resistant pathogens like MRSA.[21]

---

19. See "Liquid Fuel for Electric Cars," *Scientific America*, December 2011.

20. http://www.foresight.org/nanodot/?p=4851 and http://www.nature.com/news/2005/051021/full/news051017-18.html

21. "Nano-size Germ Killers," *Scientific American*, December 2011.

Not only cars have gotten smaller. In 2007, state-of-the-art commercial integrated circuits had *features* sized 45 nm across. In 2012, Intel will be mass-producing entire *transistors* of only 20 nm. Engineering projections show that Intel's newest circuit technology (and also a competing approach favored by other companies in the semiconductor industry) will scale down to 7-nm transistors.[22]

In a recent interview,[23] Foresight Institute president Christine Peterson declared the nanomaterials phase established and the nanodevice phase underway. She concluded the interview by summarizing the range of predictions, some as optimistic as ten years from now, for the availability of complex nanomachinery: systems of nanodevices. Her consensus guesstimate of twenty-five years would reflect slower progress than forecast (using slightly different terminology) in the 2007 roadmap. The challenges may be economic more than technological—attributable, as Ms. Peterson alludes, to the recession that began soon after the roadmap and the slow recovery that followed, and their continuing impact upon R & D budgets.

---

22. "Transistor Wars: Rival architectures face off in a bid to keep Moore's Law alive", http://spectrum.ieee.org/semiconductors/devices/transistor-wars/0

23. http://www.mypaper.se/show/semcon/show.asp?pid=345278821966177 (and within the magazine issue, the article "Infinite Nanotech Possibilities."

### *For further reading*

The National Nanotechnology Initiative: http://www.nano.gov/

National Institutes of Health: http://nihroadmap.nih.gov/nanomedicine/

Department of Energy: http://web.archive.org/web/20101215111122/http://energy.gov/sciencetech/nanotechnology.htm

The Foresight Institute: http://www.foresight.org/

The Institute for Molecular Manufacturing: http://www.imm.org/

Nanotechnology Now (general news): http://www.nanotech-now.com/

# *Our Place in the Universe*

Up to giga-light-years and down to nanoscale … we've covered a lot of territory. But what does it mean? What is humanity's place in the universe?

Is it all "A Matter of Perspective?" Or is our lot one of "Insignificance?"

You decide.

# *A Matter of Perspective*

"… Eventually reducing projected annual auto emissions by perhaps as much as five percent. The senator also indicated that her bill would …"

Hugh Robinson angrily twisted the short-wave radio knob to off. *Too little and far too late*, thought the Greenshield activist. Piddling improvements in cars would not begin to compensate for toxic wastes, ozone depletion, topsoil erosion, Frankenfoods, fertilizer runoff, or—he paused to savor the lush foliage all around him—despoliation of the tropical rain forest.

That last problem, however, was one he *did* know how to address.

◆❖◆

Gaea was slow to notice the eruption of humanity. Within a few tens of revolutions, a mere blink of the planetary sensorium, the primates had changed from scattered symbionts to a pandemic. In their teeming billions, no land seemed immune from them; their wastes choked the oceans and fouled the air.

*Too bad*, she thought. *They had been off to a promising start.* Perhaps a small environmental tweak would help.

In the bowels of the Earth, great currents of magma roiled. Several revolutions of effort would be needed to fully overcome inertia and viscosity, but then, as the flow of molten metal altered, the global magnetic field would realign. The upper northern latitudes, site of the most nettlesome human infestations, would experience a major thinning of the ozone layer; far more ionizing radiation would reach the surface. Mortality, and perhaps mutation, would increase.

Committed to a course of action, Gaea experienced satisfaction. A die-off every now and then was good for a species.

◆❖◆

As the planets continued their stately dance, Aten pondered its slow, deep thoughts. Great masses of the solar consciousness welled up from its fluid depths, sometimes spreading sedately across the luminous surface, sometimes leaping with great joy far into the vacuum before falling back, spent.

Pale glimmers of information fell inward from the void, data often difficult to discern against the glorious background of its own nuclear fire. Over several sunspot cycles, a picture emerged. The third world outward once more showed disconcerting signs of degradation and instability.

Stellar thoughts turned dark. The second planet had gone this way, megacycles earlier. Perpetually shrouded, that world was now beyond even Aten's ability to fix. But not yet the third planet .... The sun still knew ways to influence events there.

Melancholy gave way to resolution. Flare after flare erupted, arching far into space. Volatiles boiled from the suddenly hot surface of a near-perihelion asteroid, slightly altering its eccentric orbit. Three more circuits, with a gravitational assist from the ringed sixth world, and the chosen asteroid would assume the desired trajectory.

A sunspot cycle or two with its atmosphere darkened by asteroid dust should cool down whatever was troubling the third world.

◆❖◆

From aeons and locations unimaginably distant, photons plunged into the maw of the galactic black hole.

Deep within the event horizon, the galactic sentience patiently assimilated the influx. From the flood of seemingly meaningless noise, #@%_^# gradually extracted the modulated quantum emissions of its peers. The fellowship of the galactic black holes was methodical but profound.

Almost three galactic rotations previously, #@%_^# had initiated an exchange with a neighbor. A burst of noise now obscured much of the long-awaited answer. The galactic sentience studied the interruption for regularities that might enable recovery of the message. The pattern, when it did emerge, was ugly.

The in-falling coruscations of light revealed the reemergence all around it of self-motivated stars and star clusters, all too many of them accompanied by planetary infections. The concept, though disgusting, was bearable—so long as the tiny creatures remained quiet. Their interruption of the cosmos-spanning dialogue, however, was beyond all endurance.

#@%_^# contemplated slowly, as did all of its kind. A substantial fraction of a galactic rotation later, its ruminations completed, it acted. Expending a few millennia of mass-energy influx, it modulated a blast of quantum radiation, a galactic shout, at its event horizon. The expanding wave would trigger a chain reaction of supernova events to purify the nearest nebulae. The supernovae would, in turn, release a much amplified flood of cosmic rays, sterilizing planetary vermin throughout a much greater volume.

*That* should improve the neighborhood decorum.

◆❖◆

(...) paused in its deliberations, distracted by it knew not what. The distraction, whatever it was, defied immediate identification. It also refused to fade away.

Puzzled and annoyed, the creature scanned a progression of dimensionalities and continua, finally pinpointing the vexation as a series of annoying galactic perturbations within a low-order universe.

Localization did not make the interruptions any less bothersome. After the passage of something analogous to, if far more complex than, time, the multiverse creature decided to damp out the disturbance. Slightly and subtly realigning its corpus, (...) increased the fraction of dark matter that resided within the offending space-time continuum.

The situation addressed, (...) resumed its work. Serenity would return soon enough, as the newly augmented universe moved swiftly through the cleansing of a Big Gnab event.

◆❖◆

The tropical breeze gently ruffling his hair, Hugh Robinson eyed his handheld radio transmitter. He would have only a few hours after releasing the tailored, aerosol-born Ebola virus to return safely to the airport. As for the unsuspecting local residents …

He took a moment to admire towering trees beneath an impossibly blue sky dotted with scudding clouds. Unseen birds sang magnificently—until a passing jetliner obliterated their calls and Hugh's lingering doubts. He pressed firmly the large red transmit button. *All* this must be protected, whatever the cost.

After all, one had to keep one's eyes on The Big Picture.

# *Insignificance*

In the beginning we were at the center of the universe—
Special.
Or so we saw ourselves until curiosity set in.
Then we discovered:
That the Earth is merely one planet among many, orbiting the sun.
That the sun is but an unexceptional star,
Born in obscurity to die in obscurity in a remote fringe of the Milky Way.
That the Milky Way is but one galaxy, unremarkable, among countless myriads.
That the universe is thirteen billion years old, give or take,
That it did fine without us for almost all of its existence,
And it will go on, unperturbed, if we leave.
We discovered all that. Ephemeral, Johnny-come-lately, back-end-of-nowhere us.
It seems we are special, after all.

*Appendix:*
*Mapping of Themes to Novels and Collections*

| | SETI, Communi-cations, and First Contact | Computer Science and Artificial Intelligence | Far-Out Physics (including FTL) | Near-Earth Space Travel | Nanotech |
|---|---|---|---|---|---|
| **Probe (1991)** | P | P | | S | |
| **Moonstruck (2005)** | P | | S | P | P |
| **Creative Destruction (2006) (*)** | S | P | | S | |
| **Fleet of Worlds series (2007-12) (#)** | S | S | P | | S |
| **Fools' Experiments (2008)** | | P | | | |
| **Small Miracles (2009)** | | S | | | P |
| **InterstellarNet series (2010)** | P | P | S | S | S |
| **Countdown to Armageddon (2010) (!)** | | | P | | |
| **A Stranger in Paradise (2010) (*) (!)** | | S | S | S | S |
| **Energized (2012)** | | S | | P | |

**Key**

(*) Short-fiction collection

(#) Written with Larry Niven

(!) As ebooks, packaged separately. In print, packaged as one volume (*Countdown to Armageddon / A Stranger in Paradise*)

P Primary theme

S Secondary theme (in a novel); story (within a collection)

## *About the Author*

EDWARD M. LERNER worked in high tech for thirty years, as everything from engineer to senior vice president, often while writing science fiction as a hobby. Since 2004 he has written full-time, and his novels run the gamut from near-future technothrillers, like *Small Miracles* and *Energized*, to traditional SF, like the InterstellarNet series. Collaborating with Larry Niven, Ed also wrote the *Fleet of Worlds* series of *Ringworld* companion novels.

Ed's short fiction has appeared in anthologies, collections, and many of the usual SF magazines. As highlighted in this collection, he also writes the occasional nonfiction technology article.

Lerner lives in Virginia with his wife, Ruth.

His website is www.edwardmlerner.com.

## *Also by Edward M. Lerner from FoxAcre Press*

### InterstellarNet: Origins

***When the First Call from the Stars Comes, Do We Even Dare to Answer?***

Life changes for everyone in general—and for physicist Dean Matthews in particular—when astronomers detect a radio signal from a nearby star. First Contact forces humanity to face hard questions, and do it fast. Every answer spawns new questions. Every solution sets in motion a new and more daunting crisis to challenge Dean, his family—and an expanding number of interstellar civilizations—for generations to come.

"One of the most original, believable, thoroughly thought-out, and utterly fascinating visions ever of what interstellar contact might really be like."
*—Dr. Stanley Schmidt, Editor, Analog Magazine*

---

### InterstellarNet: New Order

***Good fences, said the poet, make good neighbors ... and interstellar distances made very good fences. Or so we thought ....***

Earth and its interstellar neighbors have been in radio contact for a century and a half. A vigorous commerce in intellectual property has accelerated technical progress for all the species involved. Ideas, riding on radio waves, routinely cross interstellar space—almost like neighbors chatting over the interstellar back fence. But there is a way over, or under, or around, almost any fence. Sooner or later, when we least expect it, the neighbors, friendly or otherwise, are going to pay a call.... *InterstellarNet: New Order* chronicles the startling events of *Second* Contact, upfront and personal, as humanity discovers that meeting aliens face to face is very different—and a lot more dangerous—than sending and receiving messages.

## Moonstruck

***Will our first contact with aliens be the dawn of a new tomorrow—or the last act in human history?***

The moon has suddenly acquired its own satellite: a two-mile-across starship that represents a hitherto unsuspected Galactic Commonwealth.

The F'thk, a vaguely centaur-like member species for whom Earth's ecology is hospitable, have been sent to evaluate humanity for prospective membership. The F'thk are overtly friendly but very private—"Information is a trade good" could almost be their mantra. The Galactics' arrival may signify the start of a glorious new era, or it may presage the cataclysmic end of human civilization. Which outcome do the aliens really desire ... and what will they do if humanity refuses to play its assigned role?

---

## Probe

***What if First Contact doesn't come the way we expect it?***

Bob Hanson, the chief scientist of a major aerospace corporation, has made an incredible discovery: a wrecked alien spacecraft adrift in the Asteroid Belt. The military enthusiastically embraces an investigation of the extraterrestrials, remarkably indifferent to the inconsistencies that begin to appear. But Hanson keeps digging ... and finds much more than he had ever bargained for. Are aliens manipulating events on Earth? Did unscrupulous corporate executives invent the aliens in search of giga-buck government contracts? Has the Pentagon fabricated an alien menace for its own purposes? Or is the truth something really unimaginable?

CPSIA information can be obtained at www.ICGtesting.com
Printed in the USA
BVOW021838270312

286213BV00001B/80/P

9 781936 771370